Fernando Acosta Urrea

Influence of elevated temperatures up to 100 °C on the mechanical properties of concrete

Karlsruher Reihe

Massivbau
Baustofftechnologie
Materialprüfung

Heft 84

Institut für Massivbau und Baustofftechnologie
Materialprüfungs- und Forschungsanstalt, MPA Karlsruhe

Prof. Dr.-Ing. Frank Dehn
Prof. Dr.-Ing. Lothar Stempniewski

Influence of elevated temperatures up to 100 °C on the mechanical properties of concrete

by

Fernando Acosta Urrea

KIT Scientific Publishing

Karlsruher Institut für Technologie
Institut für Massivbau und Baustofftechnologie

Influence of elevated temperatures up to 100 °C on the mechanical properties of concrete

Zur Erlangung des akademischen Grades eines Doktor-Ingenieurs von der KIT-Fakultät für Bauingenieur-, Geo- und Umweltwissenschaften des Karlsruher Instituts für Technologie (KIT) genehmigte Dissertation

von Fernando Acosta Urrea M.Sc. aus Armenia, Kolumbien

Tag der mündlichen Prüfung: 24. Juli 2017
Referent: Prof. Dr.-Ing. Harald S. Müller
Korreferent: Prof. Dr.-Ing. Harald Budelmann

Impressum

KIT Scientific Publishing

Karlsruher Institut für Technologie (KIT)
KIT Scientific Publishing
Straße am Forum 2
D-76131 Karlsruhe

KIT Scientific Publishing is a registered trademark of Karlsruhe Institute of Technology.
Reprint using the book cover is not allowed.

www.ksp.kit.edu

Print on Demand 2018 – Gedruckt auf FSC-zertifiziertem Papier

ISSN 1869-912X
ISBN 978-3-7315-0795-6
DOI 10.5445/KSP/1000082992

Abstract

The main goal of this thesis was to develop a physically based material model capable of predicting the influence of elevated temperatures up to 100 °C on the mechanical properties of concrete. The motivation resulted from the necessity of assessing the effects of ageing on the mechanical properties of concrete in order to determine the structural integrity of buildings exposed to ambient environmental conditions.

The mechanical properties considered by the model are strength and stiffness, by means of compressive strength, tensile strength and modulus of elasticity and time dependent deformations, covering creep and shrinkage. This document summarises the process of evolution and achievement of the results of the investigation. The work is written according to the time sequence followed along the investigation. It starts with a literature review, followed by the planning and conduction of a series of experiments, which served to generate the required knowledge and data that were finally used to develop the material model.

Before the planning of the experiments, a literature review aimed at identifying the most important features playing a role on the mechanical behaviour of concrete subject to elevated temperatures was conducted. The literature review made clear that it is not just temperature that influences the mechanical properties of concrete. Besides temperature, the moisture content is also an important factor. Open questions are related to their interaction with respect to the mechanical behaviour of concrete. In consequence, it was necessary to identify how these two factors influence each other and to what extent their impacts on the mechanical properties of concrete can be considered independent from one another. For concretes made with quartzitic aggregates, at elevated temperatures and high moisture content, hydrothermal reactions between silicone dioxide from the aggregates and calcium hydroxide from the cement paste can take place, forming new calcium silicate hydrate phases. These reactions represent a third factor affecting the mechanical behaviour of concrete, but according to the consulted sources, they do not influence the mechanical properties of concrete substantially.

The experimental program was planned and conducted based on the literature review focusing on the open scientific questions. The experiments were aimed at characterizing the effects of moisture content and temperature on the mechanical properties of concrete. It was assumed that these effects could be appraised separately and in case hydrothermal reactions would occur, their influence on the mechanical properties could be neglected. Samples of three concrete mixtures were conditioned at different relative humidities and subject to diverse room temperatures. Tests of compressive strength, tensile strength, modulus of elasticity, creep and shrinkage were carried out. The relative humidity in the concrete pores was measured in several samples throughout the whole process in order to register the moisture content of the samples that were tested.

From the experimental results it can be stated that the influence of hydrothermal reactions on the concrete compressive and tensile strength cannot be neglected. Nevertheless, in accordance with the literature review, it was found that these reactions occurred only at very high contents of water in the concrete microstructure at temperatures higher than approx. 60 °C.

The results of the experiments were rigorously analysed. This analysis led to the formulation of physically based mathematical expressions to describe the mechanical behaviour that the concrete samples showed in the experiments. Formulations to predict the influence of moisture content and temperature on the concrete strength and stiffness as well as creep and shrinkage were developed based on the test results. The formulations showed a very good compliance with the measurements, however, their applicability is limited because the relative humidity of the concrete must be known in advance. However, if not measured, this may be estimated from formulas based on diffusion theory.

The final part of the thesis deals with the development of an overall material model to predict the mechanical behaviour of concrete subject to elevated temperatures up to 100 °C. The conception of the material model represents an effort to improve the applicability of the developed formulations.

Due to the fact that the moisture content of concrete can vary with time, the only way to implement the formulations in a material model was to couple them with a model of moisture transport in concrete. Using the large amount of measurements of relative humidity conducted on the concrete samples during the experimental program, it was possible to conceive and calibrate a model of moisture transport for concrete. The model of moisture transport was developed based on the theory of diffusion. The combination of the moisture transport model with the formulations to predict the mechanical properties of concrete resulted in a new overall material model with the capacity to calculate the time dependent behaviour of the concrete mechanical properties subject to elevated temperatures and variable moisture content. This model was tested by comparing it with the experimental results showing a very good accuracy. Moreover, the model is capable of describing the time dependent behaviour of the mechanical properties based on physically sound assumptions which allows to extend its applicability beyond the concretes treated in the investigation.

Preface

The present thesis summarises my work at the Institute of Concrete Structures and Building Materials (IMB) of the Karlsruhe Institute of Technology (KIT), Germany, during the period 2011 - 2016. The dissertation deals with the experimental determination of material parameters and the development of a new overall material model capable of describing some key aspects of the concrete mechanical behaviour under the influence of elevated temperatures up to 100 °C.

I would like to express my gratitude to many people who contributed to the accomplishment of this work. First and foremost, I am greatly indebted to my thesis supervisor, Prof. Harald S. Müller. I thank him for his continuous interest in this work, for supporting this project and giving such thoughtful feedback, always aimed at moving me forward. Dr. Michael Haist was supportive of this and many other endeavours during my time at IMB. His inspiration, rigorousness and scrutinising spirit are undoubtedly virtues worth to be emulated. They have both had an immense impact on my personal and academic evolution, and for that I offer my heartfelt thanks.

I am profoundly grateful to Prof. Arcesio Lizcano who gave me the opportunity to come to Germany and gather new experiences. My former boss, Dr. Peter M. Mayer for making possible my entrance into the IMB. They both believed in me and paved the path I have been following in my professional development.

I want to express my gratitude to my colleagues Ferdinand Borschnek, Martin Umminger, Marco Kromer, Astrid Hirsch, Raphael Breiner, Julian Link, Jack Moffatt, Zorana Djuric, Andreas Wiedmann, the scientific staff of IMB , Dr. Michael Vogel, Dr. Engin Kotan, Dr. Viktoria Malarics-Pfaff, and the administrative staff integrated by Frau Neugebauer, and Frau Schmitt for their kindness and their various forms of support. My thanks are also extended to the staff of the Material Testing and Research Institute (MPA) for supporting me during the preparation and conduction of the experiments.

The financial support of this research came from the German Federal Ministry of Education and Research (BMBF) in the framework of the research project AURIS as part of the programme "Research for Civil Security" and German government's high-tech strategy. This support is gratefully acknowledged herewith.

I thank my parents Alvaro and Clara Ines and my brothers Alvaro and Alejandro for their love and wholehearted support. Last but not least, I would like to thank and dedicate this work to my wife Melisa and our baby boy Agustin, her love, her patient and the immeasurable encourage she gave me made the conclusion of this odyssey possible.

Stuttgart, March 2018

Contents

Notation

CSH	Calcium silicate hydrate
CH	Calcium hydroxide
w/c-ratio	Ratio of mass of water to mass of cement [-]
p	Partial pressure of water vapour [Pa]
p_{sat}	Saturation pressure of water vapour [Pa]
v	Adsorbed volume of gas [m^3]
v_m	Volume of gas when a unimolecular layer is adsorbed [m^3]
C	Constant of the BET equation [-]
σ	Surface tension of water [N/m]
θ	Contact angle between the surfaces of liquid and solid [°]
r_c	Capillary radius [m]
ρ_w	Density of water [kg/m^3]
R_D	Specific gas constant of water vapour [J/(kgK)]
RH	Relative humidity of the environment [%]
T	Temperature [°C]
$\mathrm{T_{ref}}$	Temperature of reference [°C]
K	Hygrothermic coefficient of concrete [K^{-1}]
h	Relative humidity in the concrete pores [-] or [%]
h_{mean}	Mean relative humidity of the concrete sample [-] or [%]
u	Moisture content in the concrete pores [%]
$\mathrm{K_{ref}}$	Permeability coefficient of concrete at reference conditions [m^2]
$\mathrm{K}_{\mathrm{T},h}$	Permeability coefficient of concrete after being submitted to a given temperature and relative humidity [m^2]
f_{c}	Compressive strength of concrete [N/mm^2]
f_{ct}	Tensile strength of concrete [N/mm^2]
E_{c}	Modulus of elasticity of concrete [N/mm^2]
D	Diffusion coefficient [mm^2/d]
D_1	Diffusion coefficient at reference conditions [mm^2/d]
α_0, h_c, n	Parameter of the function f(h) [-]
Q	Activation energy of Diffusion [J/kg]
r	Radius [mm]
h_∞	Relative humidity of the environment [-]
h_0	Initial relative humidity of the concrete pores [-]
r_0	Constant to account for units of length oder than mm

a, b, c	Parameters to calculate the hygrothermic coefficient K according to the w/c-ratio of the concrete mixture [-]
D_{f_c}	Effect of drying on the compressive strength of concrete [-]
$D_{f_{ct}}$	Effect of drying on the on the tensile strength of concrete [-]
D_{E_c}	Effect of drying on the on the modulus of elasticity of concrete [-]
S_{f_c}	Effect of thermal incompatibility on the compressive strength of concrete [-]
$S_{f_{ct}}$	Effect of thermal incompatibility on the tensile strength of concrete [-]
S_{E_c}	Effect of thermal incompatibility on the modulus of elasticity of concrete [-]
h_T	Relative humidity of the concrete pores at a given temperature T [-]
$h_{T_{\mathrm{ref}}}$	Relative humidity of the concrete pores at reference temperature T_{ref} [-]
a_{f_c}, $a_{f_{ct}}$, a_{E_c}	Parameters to account for the influence of the relative humidity on the factors of thermal incompatiblity [-]
ε_{c}	Total deformation of concrete [-]
$\varepsilon_{\mathrm{cc}}$	Creep deformation of concrete [-]
$\varepsilon_{\mathrm{cs}}$	Total shrinkage deformation of concrete [-]
$\varepsilon_{\mathrm{cds}}$	Drying shrinkage deformation of concrete [-]
K_{cds}	Factor of correlation between changes in the concrete relative humidity and drying shrinkage [-]
$K_{\mathrm{cds}T}$	Factor to account for the temperature dependency of the drying creep [-]
φ	Creep coefficient [-]
φ_{bc}	Basic creep coefficient [-]
φ_{dc}	Drying creep coefficient [-]
$\beta_{bc0}(f_{cm})$	Factor to account for f_{c} on the calculation of basic creep according to MC 2010 [-]
$\beta_{bc}(t,t_0)$	Factor to calculate basic creep according to MC 2010 [-]
K_{bc1}, K_{bc2}	Factors to calculate the basic creep coefficient [-]
φ_{dc0}	Parameter to calculate drying creep according to EN 1992-2 [-]
K_{dc}	Factor to calculate the drying creep coefficient [-]
$K_{\mathrm{bc}T}$	Factor to account for the temperature dependency of the basic creep coefficient [-]
$a_{K_{\mathrm{bcT}}}$, $b_{K_{\mathrm{bcT}}}$	Temperature-dependent parameters of K_{bcT} [-]
K_{dcT}	Factor to account for the temperature dependency of the drying creep coefficient [-]

Chapter 1

Introduction

1.1 Problem statement and objectives of the investigation

In the last years the estimation of the service life of concrete structures is becoming increasingly more important. Mostly due to economical but also to environmental and political reasons, many concrete structures are being used beyond the service life they were designed for. Thus, nowadays the necessary repair of damage and maintenance of concrete structures concerns many private and public institutions. In order to determine whether a given concrete structure maintains its structural integrity, safety evaluations based on models, monitoring and inspections must be periodically conducted. Performing safety evaluations requires an assessment of the effects of concrete ageing because the concrete properties vary in time according to the environment to which the concrete is exposed. The vast majority of the concrete structures are exposed to ambient environmental conditions in which variation of the moisture content is the most important factor contributing to the ageing process of the concrete being constituted by hydration and/or some kind of polymerisation process. Yet, concrete structures used in the energy industry such as cooling towers or buildings for thermal energy storage are additionally subject to elevated temperatures. These buildings belong to the group of sensitive infrastructure whose vulnerability is decisive for human civilization.

Between the years 2011 and 2015 the German Federal Ministry of Education and Research (BMBF) supported the project AURIS as part of the program "Research for Civil Security" of the German government's high-tech strategy. AURIS was a research project aiming at developing an autonomous risk and information system for structural analysis and health monitoring of security-relevant buildings. This safety management system consists essentially of a measuring system to evaluate the condition of the building structure before, during and after exposure to an extraordinary load with the purpose of protecting people in buildings of critical infrastructure against extraordinary events, e.g. during an explosion, accident or earthquake.

An aged building responds differently to a sudden change in loads than a newly constructed building. On this background, the Institute of Concrete Structures and Building Materials (IMB) of the Karlsruhe Institute of Technology (KIT), Department of

Building Materials was entrusted under the joint research project AURIS with the conduction of the subproject "Development of a state model for concrete structural elements". The main objective of this subproject was the development of material models for concrete to predict the bearing capacity of a concrete structure during its service life and at the moment an extraordinary event occurs. The objective of the project was achieved by conducting systematic investigations on concrete samples subject to elevated temperature and various ambient humidities.

Later, a more ambitious goal constituting the main objective of this thesis emerged from the results and expertise gathered with the research project. This consisted in the development of a new overall material model capable of predicting the mechanical behaviour of concrete subject to elevated temperatures up to 100 °C. The new material model estimates the time, temperature and moisture dependency of the compressive strength, tensile strength, modulus of elasticity, creep and shrinkage of concrete. Consequently, this thesis provides specific knowledge about the influences of temperature and moisture content on the mechanical behaviour of concrete as well as a tool to calculate these influences. This shall contribute to the field of investigation of estimating the structural safety under ordinary and extraordinary loadings as well as to the area of evaluating the service life of concrete structures.

1.2 Structure of the document

The thesis is structured in four main parts: literature review, experimental program, conception of models and conclusions. The concrete mechanical properties throughout the document are grouped in two categories. These are the concrete strength and stiffness including compressive strength, tensile strength and modulus of elasticity and the concrete time dependent deformations referring to creep and shrinkage.

Following this introduction, in Chapter 2 a literature review embraces the current knowledge about the influence of temperature on the mechanical properties of concrete. This chapter served to identify the most important factors playing a role in the relations between concrete mechanical properties and temperature and allowed the design of an ambitious and well-oriented experimental program.

The planning and results of the experimental program are presented in the two following chapters. The production and storage of the concrete samples and the description of the conducted tests are presented in Chapter 3, while Chapter 4 contains the execution of the experiments accompanied by a rigorous analysis and discussion of the results.

In Chapter 5 the experimental results are used to formulate mathematical expressions to relate the influence of temperature and moisture content with changes of the concrete mechanical properties. These formulations are then implemented in the material model presented in Chapter 6.

The work is finally summarized in Chapters 7 and 8 comprising the most important results and conclusions of the investigation and an outlook on the issues that remain unsolved including those that arose during the investigation.

Chapter 2

Literature review

This chapter aims at providing an overview of the various agents that contribute to the mechanical behaviour of concrete and their interaction with temperature. The literature review begins by asserting the importance of water in the formation of the concrete microstructure, its presence in the hardened concrete, and its interaction with the concrete components and the environment. The mechanisms associated with the process of movement of water within the concrete microstructure are briefly introduced including some theoretical approaches to describe them. Additionally, the chapter presents a compilation of experimental results from previous investigations concerning the effects of temperature on the mechanical properties of concrete. Finally, own considerations based in the literature review are presented which guide the design of the experimental investigations to be conducted.

2.1 Role of water in the concrete microstructure

Water always plays an important role on the life of concrete. Right from the very beginning, as a component of the fresh concrete, the amount of water used in the concrete mixture governs its workability [64]. During hydration, the developed hydrates fill the space between the solids seizing the space originally occupied by water. The water not consumed by the hydration reaction remains in the pore space defining thereby the microstructure of the hardened cement paste, i.e. porosity of the produced material. The lower the porosity the higher the strength and durability of the concrete [64, 116]. Moreover, during the hydration reaction empty capillary porosity within the cement paste component of the concrete is created due to chemical shrinkage [61, 115, 149], as well as to the loss of water to the ambient enviroment. The kinetics of the hydration reaction are linked to the presence of water. An empty pore space created due to drying or self-desiccation will not be filled with hydration products, because due to the lack of water the curing process slows down and terminates at a lower degree of hydration than the one that could be achieved under saturated conditions [21].

The role of water goes beyond its importance for the workability of concrete and the hydration reaction of the cement paste. On the basis of a fully hydrated concrete with a constituted microstructure, the mechanical properties of the concrete will still undergo considerable changes if its water content varies. According to Dahms [41] the compressive strength of concrete increases up to around 30 % after drying compared to the same concrete in saturated conditions. Bonzel and Kadleček [26] reported a decrease of the tensile strength of concrete shortly after subjecting it to drying and a continuously increase of the tensile strength afterwards. Pihlajavaara [112], after being engaged for over 10 years in the investigation of the ageing phenomena of drying in concrete, came to the conclusion that the drying causes strengthening and the wetting weakening of the hardened cement paste.

Also the deformation behaviour of concrete shows a significant dependency on water content. Concretes under saturated conditions have a higher modulus of elasticity than dry concretes, because the water in the pores is incompressible and cannot scape so quickly when the concrete is subject to rapid loading [64]. Both creep and shrinkage deformations are the higher, the higher the water content and the moisture gradient within the concrete member [98].

2.1.1 Structure of hardened Portland cement paste

Knowing the nature of the hardened cement paste is essential for the understanding of the processes that can lead to changes on the mechanical properties of hardened concrete. Henceforth, the essential characteristics of the composition and structure of the hardened cement paste are following summarized.

2.1.1.1 Microstructure, porosity and pore water

The Portland cement paste, in its fresh state, is composed of cement grains in an aqueous solution. As cement grains react with water, new solid phases are produced. These solid phases consist of a dense, though porous substance, called cement gel and other products playing a relative minor role as structural units, such as crystalline calcium hydroxide (CH), ettringite, etc. The cement gel is composed of gel particles consisting primarly of calcium silicate hydrates (CSH) of various compositions with a large specific internal surface area of about 700 m^2 per cm^3 of solid [117], and interstices among those particles, called gel pores. In case the cement gel does not fill all the spaces within a specimen of cement paste originally occupied by water, capillary porosity is produced.

According to Powers [117], the capillary porosity exists in the hardened cement paste as interconnected channels or, if the structure is dense enough, as cavities interconnected only by gel pores. The capillary porosity depends on the original proportion of water in the cement paste, which is usually expressed as the ratio of mass of water to mass of cement (w/c-ratio) in the original mixture. During the hydration reaction the cement gel is produced in water-filled capillary cavities, and when all those cavities become full, no further hydration of cement can occur. In order to hydrate all the cement grains in a given cement paste, the mass of water added must be of approximately 40 to 43 % the mass of cement (0.40 $\leq$ w/c-ratio $\leq$ 0.43). For w/c-ratios lower than 0.40 not all the available cement grains will be able to react with water, leaving some unhydrated cement grains in the cement paste, while the excess of water by w/c-ratios higher than 0.40 leads to the formation of capillary pores (see Fig. 2.1). Hence, the w/c-ratio has a substantial influence on the development of the cement gel microstructure and determines the physical properties of the hardened cement paste.

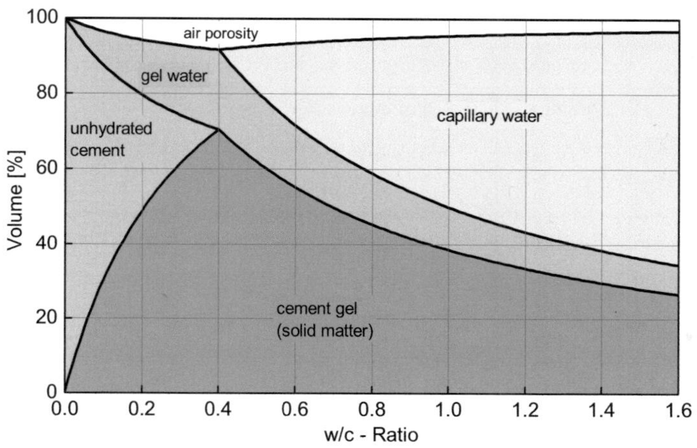

Figure 2.1: Volume proportions in relation to w/c-ratio in fully hydrated hardened cement paste [40]

Besides the microscopic gel and capillary porosity also macroscopic air and compaction voids are presented in the hardened cement paste. The subdivision and nomenclature of the pores are not uniformly managed in the literature and depend on the arbitrary definition of the respective authors. In a general way the IUPAC[1] [N18] classifies the pores of a porous solid according to their sizes: micropores have diameters less than about 2 nm, mesopores have diameters between 2 and 50 nm and macropores have diameters larger than about 50 nm.

Other classifications specifically developed for hardened cement paste have also been proposed. Table 2.1 presents the pore classifications according to Mindess and Young and Setzer [94, 140]. These classification are based on the interaction between the internal surfaces of the hardened cement paste and water. Within the microscopic ranges the interaction between water molecules and solid matter is predominantly based on surface physical principles, while for the macroscopic voids, this interaction is related to volume and mass-based processes. The threshold value between microscopic and macroscopic lies according to Setzer [141] by a pore width of about 0.4 µm.

[1] International Union of Pure and Applied Chemistry.

Table 2.1: Classification of pore sizes in hardened cement pastes [94, 140]

	Designation	Diameter	Description	Role of water
Mindess and Young	Capillary pores	10 - 0.05 μm	Large capillaries	Bulk water
		50 ~ 10 nm	Medium capillaries	Moderate surface tension forces
	Gel pores	10 - 2.5 nm	Small (gel) capillaries	Strong surface tension forces
		2.5 ~ 0.5 nm	Micropores	Strongly adsorbed water; no menisci
		< ~ 0.5 nm	Micropores "interlayer"	Structural water involved in bonding
Setzer	Coarse	> 4 mm	Empty pores	-
	Capillary pores	< 4 mm	Macro capillaries	Macroscopic bulk water
		< 120 μm	Meso capillaries	
		< 4 μm	Micro capillaries	
	Gel pores	< 120 nm	Meso gel	Prestructured condensate
		< 4 nm	Micro gel	Structural surface water

Both classifications presented in Table 2.1 assume that the water stored in the gel pores is influenced by the surfaces of the gel particles. By considering the diameter of the gel pores and taking into account that the diameter of a single water molecule is of about 0.28 nm, it can be recognized that only a small number of water molecules can be stored within the gel pores, and therefore, all of them are located within the region of influence of the gel particle surfaces. These influences are the higher, the smaller the diameter of the gel pores.

2.1.1.2 Models of hardened Portland cement paste

In the last decades several authors have dealt with the challenge of assessing the influence of the moisture content on the mechanical properties of concrete through the conception of models of hardened cement paste. These models aim to illustrate the actual conditions that prevail in the hardened cement paste by describing, in a simplified and approximated way, its microstructure. In the following paragraphs some commonly used cement paste models are briefly summarized.

Cement paste model from Powers and Brownyard

The structure of cement paste was originally deduced by Powers and Brownyard [120] from data on water vapour adsorption isotherms. Many authors have addressed their work: well-known textbooks like those from Czernin [40], Taylor [148] and Neville [106] summarize the most important features of the model; Hansen [71] wrote an explanation of how the model was derived; and more recently Brouwers [28, 29] recapitulated their work demonstrating that its results enable the quantification of the reaction products of the four clinker phases[2].

According to Powers [117] the cement paste is composed of cement gel, crystals of calcium hydroxide (CH), some minor components, residues of the original cement, and capillary cavities. Water is presented in three different states: chemically bounded water, also known as non evaporable water, with a density of approximately 1.22 g/cm^3; physically adsorbed water, which occupies the gel pores and is influenced via surface forces by the gel particles; and free water occupying the capillary pores. The model of Powers and Brownyard is capable of explaining many physical properties of hardened cement paste and despite the fact that it was proposed over 50 years ago, it is the only available model capable of calculating the volumetric composition of the hardened Portland cement pastes quantitatively.

Powers used this conception of the cement paste microstructure as basis for the later development of a model to explain the mechanisms of creep and shrinkage of hardened cement paste [118, 119]. By means of thermodynamics, Powers described the behaviour of water in a simplified geometry of a wedge-shaped gap of solid matter containing adsorbed water and capillary condensed water (see Fig. 2.2). At a given relative humidity in the pore microstructure of the hardened cement paste, water absorbed by the free surfaces builds a water film, additionally, a meniscus forms holding capillary condensed water. According to Powers, in the narrow spot where the two adsorbed films meet (hindered adsorption), disjoining pressure[3] is developed . This pressure is compensated by the stiffness of the gel particles and van der Waals forces.

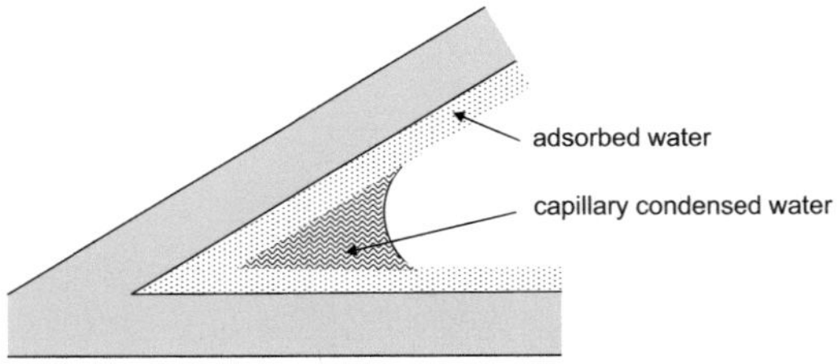

Figure 2.2: Wedge geometry according to Powers [119]

Powers explained the elementary processes of creep and shrinkage with the model presented in Fig. 2.2. The action of a force disturbs the equilibrium of the system. Initially, the water at the narrow spot must carry part of the loads (load bearing water), which causes a change on the molar free energy of the adsorbed film. The water reacts by slowly diffusing from the spot of hindered adsorption to the free adsorption areas.

[2] Alite C_3S, belite C_2S, tricalcium aluminate C_3A, tetracalcium aluminoferrite C_4AF.

[3] The concept of disjoining pressure was first formulated by Derjaguin [44], it arises when two surface layers mutually overlap.

According to Powers, when a new equilibrium is reached, the creep process ends. In the same way, the equilibrium can be disturbed by changes in the relative humidity of the pores. This eventually droved Powers to describe creep and shrinkage as two phenomena ruled by the same process of adsorption and desorption of water.

Stimulated by the general ideas of the creep mechanisms from Powers, Bažant developed the theory of hindered adsorption for concrete. This theory incorporates elements of the surface thermodynamics in the model from Powers. Bažant used the description of free adsorption proposed by Guggenheim [66], which is based on a surface phase of finite thickness and constitues an improvement of the classical surface thermodynamics from Gibbs [63] in which a zero thickness for the surface phase is assumed. Bažant developed a consistent thermodynamical formulation to describe hindered adsorption based on surface phases of finite thickness and including the action of disjoining pressure. The mathematical formulation of the theory was gradually presented in [5, 6, 7, 8], and further reviewed in a broad context in [9].

Cement paste model from Feldman and Sereda

The model from Feldman and Sereda [54, 55] was based on surfaces areas and porosities obtained by measuring N_2 adsorption in hydrated Portland cement. They emphasized on the fact that water interacts so strongly with the solid phases through chemisorption that the pore structure of the hardened cement paste is unstable in water vapour adsorption experiments, like those conducted by Powers and Brownyard to develop their model of cement paste. According to Feldman and Sereda, the hardened cement paste is composed of gel particles with a layered irregular structure. Water is not only absorbed by the gel particles, it can also enter into the layers and play a role as a structural component of the cement paste (see Fig. 2.3).

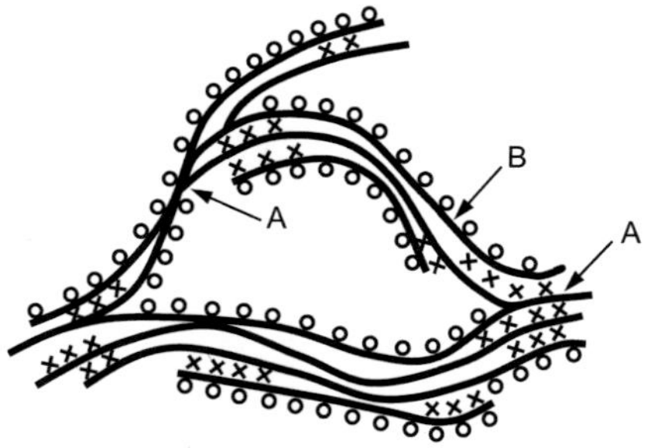

Figure 2.3: Model of cement paste structure according to Feldmann and Sereda. CSH layers (B) are represented as lines with some points of connection between them (A); crosses designate the interlayer water molecules and circles the adsorbed water molecules [54]

Depending on the temperature and relative humidity, the interlayers can contain more or less water influencing the distance between layers. In this way, the model explains the effect of vapour pressure on the modulus of elasticity and suggests that the entry and exit of interlayer water play an important role on creep [52].

Cement paste model from Wittmann and Setzer

In the model proposed by Wittmann and Setzer [143, 152] also known as the Munich model, the hardened cement paste is presented as a xerogel[4]. The model describes the influence of water on the cement paste microstructure on the basis of two humidity regions: a lower humidity region from 0 % to 40 % relative humidity (RH) in which the surface energy of the gel particles plays a fundamental role and a higher humidity region between 60 % and 100 % RH primarily influenced by disjoining pressure. Fig. 2.4 shows the expansion process undertaken by a dry hardened cement paste in the two regions of relative humidity.

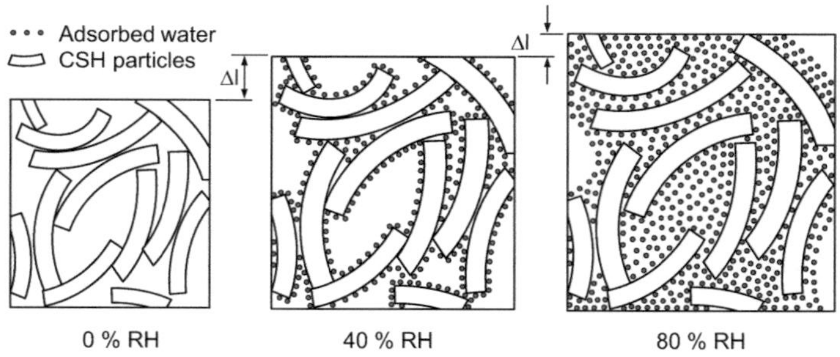

Figure 2.4: Representation of a dry hardened cement paste (left), the expansion of the cement paste in the lower RH region (centre) and the separating action of the disjoining pressure at some contact points in the higher RH region (right) [152]

In the lower RH region the model assumes that the surface free energy of the hardened cement paste maintains the gel particles together and as water molecules are adsorbed, the surface free energy decreases leading to an expansion of the gel particles. This expansion is according to Bangham and Fakhoury [15] linearly proportional to the energy changes. The first layer of adsorbed water molecules strongly influences the surface free energy of the gel particles, however, additional layers do not have much impact on the surface free energy which limits this effect to relative humidities up to 40 %. Measurements conducted by Adolphs [1] to determine surface energy relationship by means of inverse gas chromatography (IGC) supported this assumption.

To describe the processes in the higher RH region, the model uses the concept of disjoining pressure as defined by Derjaguin [44]. Based on the measurement method described by Splittgerber and Wittmann [145] and particularly on the experiments conducted by

[4] A solid formed from a gel by drying resulting in unhindered shrinkage.

Splittberger [144], it was demonstrated that, if the humidity of a system in which two plates are held together by van der Waals forces is increased, with increasing thickness of the absorbed film, a decrease of the attractive forces takes place. At a certain vapour pressure, the plates are separated by a liquid film defining the point at which the disjoining pressure of the water exceeds the attractive van der Waals forces. In general it can be stated that an increase of vapour pressure causes a separation of the surfaces that are only held together by van der Waals forces, leading the gel structure to undergo macroscopic expansion.

The model differentiates between primary bonds (chemical) and secondary bonds (van der Waals). Conforming to the Munich model, only the particles connected by secondary bonds are separated due to the action of the disjoining pressure.

2.1.2 Moisture storage in the concrete microstructure

2.1.2.1 Mechanisms regulating moisture ingress

The pore structure of the hardened cement paste is exposed to the moisture of the surrounding environment. The interaction between moisture of the surrounding air and hardened cement paste is regulated by two mechanisms.

- At low relative humidities prevails the adsorption of water vapour molecules on the pore walls. The adsorption describes a process in which the accumulation of substances on the surface of a solid takes place. The most common relation used to describe adsorption was proposed by Brunauer et al. [34] as an extension of the Langmuir's theory of monomolecular adsorption [92]. Based on multimolecular adsorption, Brunauer et al. derived an expresion to calculated the volume of adsorbed gas in a porous solid known as the BET equation (see Eq. 2.1). This equation is used in DIN ISO 9277 [N12] as standard procedure for determining the specific surface area of porous and non-porous solids.

$$\frac{p/p_{sat}}{v \cdot (1 - p/p_{sat})} = \frac{1}{v_m \cdot C} + \frac{C-1}{v_m \cdot C} \cdot p/p_{sat} \tag{2.1}$$

  In Eq. 2.1 p and p_{sat} are the partial pressure and saturation pressure of the gas, v is the volume of gas adsorbed, v_m is the volume of gas adsorbed when the entire adsorbent surface is covered with a complete unimolecular layer and C is a constant dependent on the heat of adsorption and the temperature.

- At high relative humidities $(p/p_{sat} > 0.5)$ and depending on the size of the pores, the volume of gas adsorbed by the solid does not coincide with the values given by the BET equation. The actual volume is higher because of capillary condensation. Capillary condensation is the process that follows the multilayer adsorption after the point at which pore spaces become filled with condensed liquid from the gas. Due to surface tension of the liquid, a meniscus forms at the liquid-gas interface. Over a curved liquid-vapour interface the vapour can condense by partial pressures below the saturated vapour pressure, as denoted by the law of Kelvin-Thomson [62, 74]. This law describes the relationship of the ratio of vapour pressure between a curved and a flat liquid surface according to the radius of curvature. Eq. 2.2 constitutes the most common mathematical description of the law of Kelvin-Thomson. This equation is based on a geometric simplification of the pore space. The pores are depicted as capillary tubes in which spherical menisci are formed [82].

$$\frac{p}{p_{sat}} = \exp\left(-\frac{2 \cdot \sigma \cdot \cos\theta}{r_c \cdot \rho_w \cdot R_D \cdot (T + 273)}\right) \tag{2.2}$$

In Eq. 2.2 p and p_{sat} are the partial pressure and saturation pressure of the gas, σ is the surface tension of water, θ is the contact angle, r_c is the capillary radius, ρ_w is the density of water, R_D is the gas constant for water vapour, and T is the temperature of the system.

2.1.2.2 Sorption isotherms

The amount of water stored in the concrete microstructure is characterized by the sorption isotherm of the material. The sorption isotherm indicates the relationship between the relative humidity of the environment and the equilibrium moisture content of the material at a given temperature. For hardened cement paste the sorption isotherms are typically s-shaped corresponding to the type II according to the IUPAC classification of adsorption isotherms [17]. The characteristic properties of a sorption isotherm for hygroscopic porous materials are schematically summarized in Fig. 2.5. In the top diagram of Fig. 2.5 the sorption isotherm is divided in two regions. A first region can be identified in a range between dry state and approximately 95 % RH, in which the water is bounded by adsorption. In this region, also known as sorption moisture or hygroscopic region [85], the lower range up to about 15 % RH is determined by a monomolecular adsorption of water on the solid surfaces followed by a linear increase of the volume of adsorbed water through multimolecular adsorption, which ends at about 50 % RH. The following progressively increasing range from approximately 50 % to 95 % RH is attributed to the capillary condensation. In the second region, for relative humidities above around 95 %, the sorption isotherm rises very steeply suggesting that in the macropore range, where no capillary condensation occurs, the pores are filled with unbounded water [82]. According to Künzel [85] in this region two subregions can be identified, a super-hygroscopic region corresponding to the filling of water up to free water saturation and a supersaturated region in which the filling continues until all pores are full with water reaching the maximum saturation.

There is no sharp transition between free and bound water. Usually free water is considered to be the part of moisture exhibiting physical properties similar to those of pure water. On the other hand bound water denotes that part of moisture that is bonded by sorption forces in the material, whose physical properties are significantly different from those of pure water.

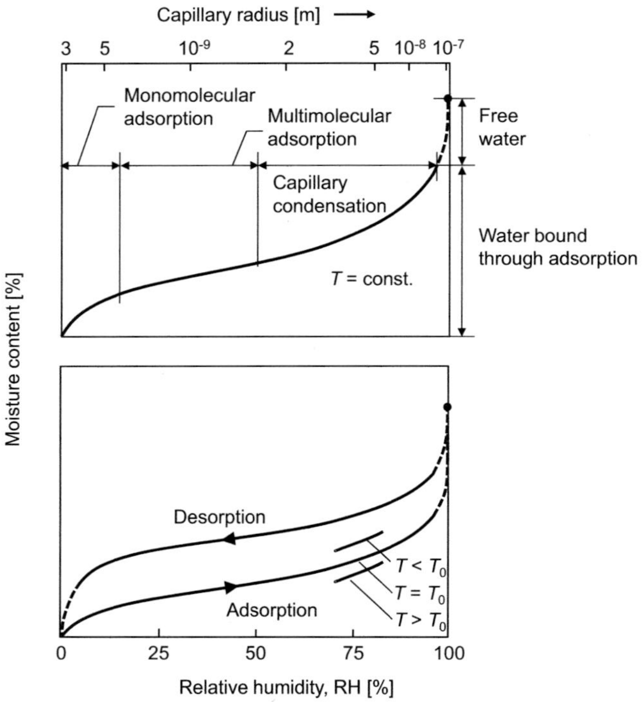

Figure 2.5: Schematic representation of a sorption isotherm for hygroscopic porous materials [82]. Top: Ranges of moisture adsorption indicating the approximated pore radius according to a modell of capillary tubes [84]. Bottom: Influence of temperature and hysteresis effect between adsorption and desorption

The bottom diagram of Fig. 2.5 depicts two important characteristics of the sorption isotherm: the influence of temperature and the hysteresis effect between adsorption and desorption. These characteristics need to be considered more precisely.

- Due to the fact that adsorption is an exothermic process [33], according to the principle of Le Chatelier[5] and the van't Hoff equation [4], it can be stated that the adsorption is hindered when temperature is increased, or rather, a temperature

[5] A system at equilibrium, when subject to a disturbance, responds in a way that tends to minimize the effect of the disturbance.

increase promotes desorption. This implies that at higher temperatures concrete reduces its capacity of storing water, the water molecules cannot be held by the internal surfaces of the concrete in the same amount when additional energy (temperature) is added to them. Therefore, for the same concrete the "warm" sorption isotherms lie under the "cold" ones. Few articles have been published regarding the effect of temperature on water sorption isotherms of cementitious materials. Nevertheless, the work from Hundt and Kantelberg [77] and more recent works from Ishida et al. [78] and Poyet [121], evidenced experimentally the impact of temperature on the sorption isotherms of hardened cement paste, mortar, and concrete. Furthermore, with the recent advent of molecular dynamic simulations, the effect of temperature on the water content within and between CSH grains can be approached numerically (see Bonnaud et al. [25]).

Another aspect that can be identified from the effect of temperature on the sorption isotherms is the fact that by constant moisture content, an increase in the temperature of the concrete will lead to an increase of the relative humidity in the pore cavities of the concrete. This can be easily visualized by drawing a horizontal line to connect two sorption isotherms at different temperatures in the bottom part of Fig. 2.5. Bažant [6, 10, 11] introduced the hygrothermic coefficient of concrete K, which describes the variation of the relative humidity dh due to a change of temperature dT at a given water content u. A simplified expression to calculate K for concrete was proposed by Bažant and Najjar [10] and is given in Eq. 2.3.

$$K = \left(\frac{dh}{dT}\right)_{u=const} = 0.0135 \cdot h \cdot \frac{1-h}{1.25-h} \tag{2.3}$$

- The hysteresis effect in sorption isotherms between adsorption and desorption has been observed at both high and low vapour pressures. Classical experiments on this issue were conducted by Powers and Brownyard [120] and Feldman and Sereda [53] but also more recent investigations from Espinosa [49] (see also Espinosa and Franke [50]), Adolphs et al. [2] and Baroghel [16] have brought more experimental evidence on this effect. Recently, an analytical approach to explain the hysteresis effect at low humidity ranges was proposed by Bažant and Bazant in a series of two papers (see Bažant and Bazant [12] and Bazant and Bažant [18]). Based on two mechanisms, they explain the hysteresis at low vapour pressure without assuming any pore collapse nor partial damage to the cement paste microstructure. Sorption and desorption in calcium silicate hydrates have also been studied using numerical molecular dynamics simulations. Works on this subject have been recently published by Bonnaud et al. [23, 24] and Brochard et al. [27].

Based on the results from an extensive experimental program, Espinosa [49] developed a method to predict the hygroscopic water content in hardened Portland cement paste and mortars at changing climatic conditions. For the development of the method which have been called the inkbottle pore method (IBP-method), a new model for cement paste based on the investigations of the structure of CSH carried out by Stark et al. [146, 147] and her own conducted experiments on sorption isotherms was deduced. The IBP-Method corresponds to the mathematical implementation of this cement paste model and takes into account the hysteresis effect of the sorption isotherms.

2.1.3 Hydrothermal effects on the microstructure of concrete

If concrete made with siliceous aggregates is subject to elevated temperatures in a very humid environment, hydrothermal reactions are expected to occur. On the surface of the quartz grains, silicone dioxide (SiO_2) dissolves in water generating silicic acids [89]. These acids react with free portlandite ($Ca(OH)_2$) from the hardened cement paste forming new calcium silicate hydrates (CSH) as shown in Fig. 2.6.

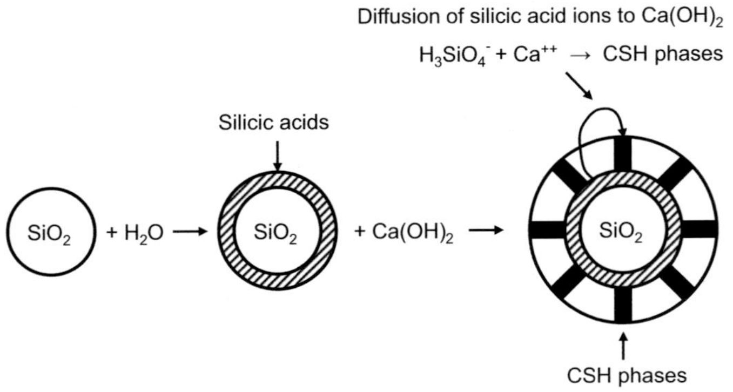

Figure 2.6: Schematic representation of hydrothermal reaction between reactive silica (SiO_2) and portlandite ($Ca(OH)_2$) [89]

The hydrothermal reactions can take place if by the given conditions of temperature and pressure both silicone dioxide and portlandite are soluble in water. The solubility of SiO_2 increases with increasing temperature up to around 340 °C [3, 56] and increasing pH value of the solution [14], it is also higher for amorphous silicone dioxide than for quartz [100, 101]. In contrast to silicone dioxide, portlandite exhibit a retrograde solubility i.e. the solubility decreases with increasing temperature [67]. Fig. 2.7 shows the solubilities of amorphous and crystalline silicone dioxides and calcium hydroxide (portlandite) in relation to the temperature. The curves are built based on the equations proposed by Gunnarson and Arnórsson [68] for silicone dioxide and the IUPAC [N17] for calcium hydroxide.

The amount of free $Ca(OH)_2$ that reacts with SiO_2 depends on the temperature and duration of the hydrothermal condition. According to Kondo [86], based on investigations conducted on mixtures of CaO and SiO_2, at temperatures below 130 °C only 30 % of the available $Ca(OH)_2$ reacts with the SiO_2 and the reaction ends substantially after 24 hours. At a temperature of 150 °C approximately 90 % of the total $Ca(OH)_2$ reacts within 24 hours. The total amount of free $Ca(OH)_2$ requires 12 hours to react with SiO_2 at 180 °C and 6 hours at 220 °C.

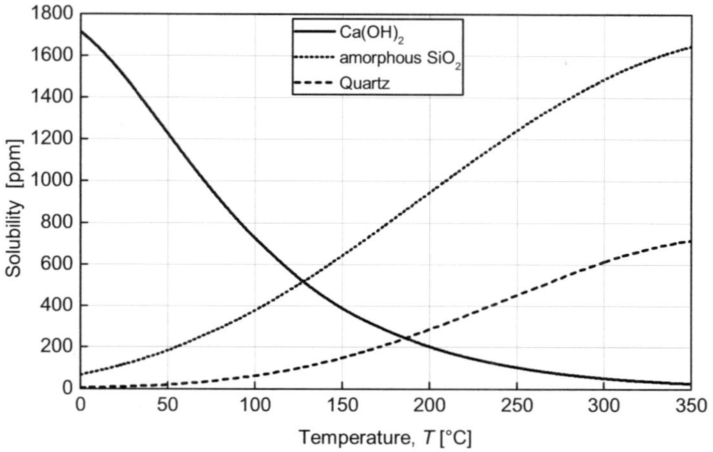

Figure 2.7: Solubility in water of calcium hydroxide, amorphous silicone dioxide, and quartz

The influence of hydrothermal reactions on the concrete properties has been studied by several authors. Seeberger et al. [138] investigated the influence of temperatures between 20 °C and 250 °C on the strength and microstructure of concretes made with siliceous and limestone aggregates. They found that due to hydrothermal reactions, concretes made with siliceous aggregates show a recovery or even an increase in the compressive strength after being exposed to elevated temperatures both for sealed and unsealed samples. Ehm et al. [47] (see also Rostásy et al. [131]) could not evidence any reaction between the free $Ca(OH)_2$ of the cement paste and the SiO_2 of the siliceous aggregates when subjecting siliceous concretes to 95 °C and 95 % RH. Results from Budelmann [35] conducted on siliceous concretes subject to 90 °C and 95 % RH show a small increment in the gel porosity which can be related to changes in the concrete microstructure due to hydrothermal reactions. Budelmann found that these changes were much more pronounced in samples heated at 90 °C under water.

For temperatures below 100 °C, probably due to the low solubility of quartz in water, the study of hydrothermal reactions in concrete has not been extensively assessed yet. However, even though the solubility is poor, below 100 °C the presence of water is guaranteed not only in a vapour state but also as fluid. Therefore, for the purpose of this investigation, a potential development of hydrothermal reactions within the concrete microstructure cannot be neglected.

2.2 Moisture transport in concrete

Moisture transport in concrete takes place due to changes in the environmental conditions. A disturbance to the equilibrium of the moisture content within a concrete member leads depending on the prevailing conditions to a moisture absorption or release. During the moisture exchange between concrete and the environment, transport of water in liquid and/or gaseous state is carried out depending primarily on the structure of the hardened cement paste [106]. There have been numerous significant publications on the moisture transport processes of porous material. They were generally first documented by the end of the 19th century and applied to concrete by the middle of the 20th. A very extensive literature review on the moisture movement and moisture properties of building materials was published in 1980 by Kießl [82] comprising 650 publications.

2.2.1 Associated mechanisms

Concerning the movement of fluids through concrete, three main processes are distinguished: permeation, diffusion and sorption. Permeation refers to flow of fluid when pressure is applied, diffusion is the movement of ions, atoms, or molecules under a differential in concentration and sorption is the capillary attraction of a liquid into empty or partially empty pores [45]. Fig. 2.8 shows an schematic overview of the main transport mechanisms in porous materials illustrating the variety and complex relationships between them.

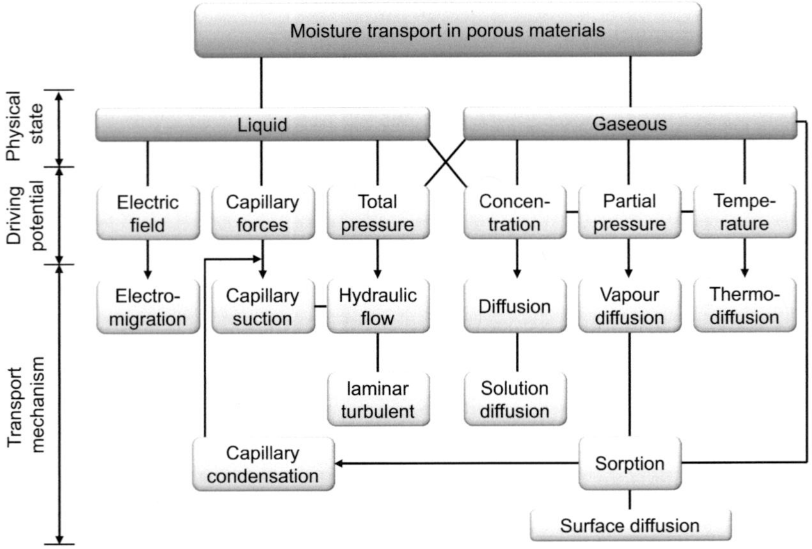

Figure 2.8: Schematic overview of the phenomena of moisture transport in porous materials [82]

Besides the three main processes named before, the fluids can also move if an electric or a temperature field is applied across the concrete. Due to an electric field the negative ions will move towards the positive electrode in a process known as electromigration [39] and through thermodiffusion the molecules of the fluid drift along a temperature gradient [46].

The dimensions of the pores are decisive to determine whether water is presented as free or bound water in liquid or gaseous state, and consequently also the mode of moisture transfer [79]. Due to the very complex geometry of the concrete microstructure, different mechanisms of moisture transport with different driving potentials perform unevenly according to the geometry of the pores in which the water is contained. Certain forms of moisture transport dominate depending on the frequency range of the pore sizes, the temperature level of the substance, and as shown in Fig. 2.5, the moisture content of the material. Rose [126, 127] described the process of wetting of a porous material in four stages according to the dominant water transport mechanisms by means of an idealized porous system consisting of a pore with a neck at each end. These stages are schematically illustrated in Fig. 2.9.

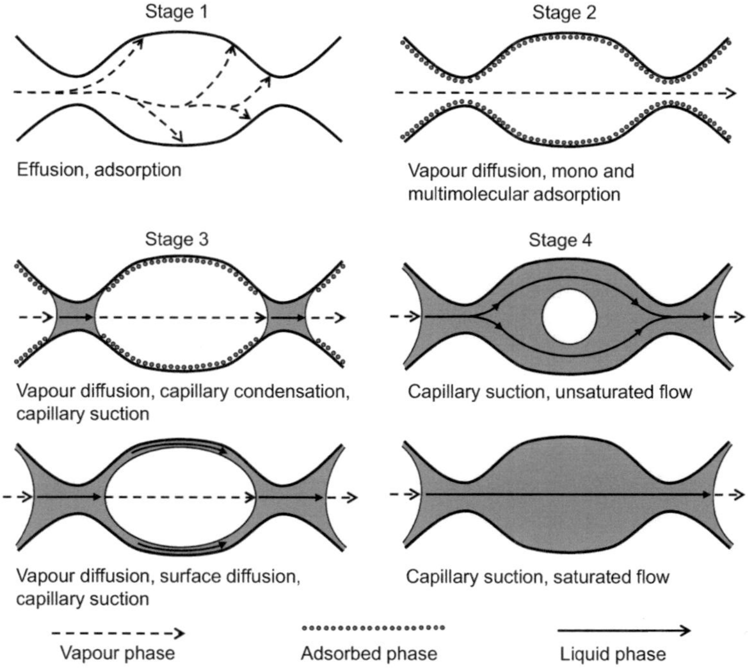

Figure 2.9: Water movement in various stages in the wetting of a porous material [127]

In the first stage, water enters the microstructure of the material not as vapour but as isolated molecules that are adsorbed by the pore surfaces. This type of transport is termed effusion or Knudsen diffusion [48]. In the second stage, unimpeded vapour transfer takes place. The water vapour moves from high to low concentration regions through diffusion while the surfaces of the pores are covered by water molecules placed in mono or multimolecular layers. In the third stage, water closes the necks shortening the effective path length for diffusion of vapour. This process is described as liquid-assisted vapour transfer. On the walls, depending on the thickness of the water film, water molecules can either be fixed to the surface through adsorption or diffuse on the surface after the film reaches a significant thickness. The fourth stage describes hydraulic flow in unsaturated or saturated conditions. The component of the fourth stage describing unsaturated conditions differs from the component of the third state where surface diffusion occurs in the fact that the air voids in stage 4 have the same curvature everywhere, which does not allow the development of concentration gradients within it, and therefore no vapour transfer takes place.

2.2.2 Describing moisture transport in porous materials

For the description of the moisture transport in porous materials a vast variety of models have been proposed. In general, one can differentiate between three different approaches to derive models for moisture transport: general linear, thermodynamic, and micromechanical approach [58].

The general linear approach has been most widely used in the literature, probably because of its simplicity and generality. According to the general linear approach, the moisture flow in porous media depends on a set of independent state variables and space-dependent material properties and their gradients. The moisture flow is calculated as the sum of the products of conductivities and gradients. Examples of the general linear approach are the models proposed by Luikov [93], de Vries [43], or Garrecht [60] which consider only the transport of moisture. Models considering moisture transport coupled with heat transfer have also been developed, like those from Krischer [90], Kießl [80], or Künzel [85]. The models differ fundamentally in the structure of the moisture transport equation which defines which transport mechanisms, physical states of the fluid, and interactions between the driving potentials and the physical states are taken into account. As the models do not necessarily use the same driving potentials, different conductivities are used and therefore, each model needs its own material parameters, which limits the comparability of the models.

The thermodynamic and micromechanical approaches are more complex and even less general. In the thermodynamic approach the entropy production is investigated. An entropy increase is responsible for the dynamics of the system. The entropy production delivers information about the moisture flow and its respective driving forces [57]. Grunewald [65] derivated the entropy production rate per volume for a heat, air moisture, and salt transport model. He concluded that the thermodynamic driving force of the diffusive water vapour flow is the gradient of the chemical potential. Hassanizadeh and Gray [75] developed a model to describe two-phase flow in porous media taking into account the effects of the surfaces separating the phases. In the micromechanical approach the heat and mass transport processes are calculated on the microscopic scale, which requires the use of a microphysical model of the porous system. Bednar [19] used micromechanics to investigate water vapour flow based on the theory of ideal gas mixtures. According to Bednar the driving potential for the water vapour flow is the vapour pressure with a small contribution of thermodiffusion. Whitaker [151] obtained the liquid water flow and liquid conductivity from the Navier-Stokes equations in the pore system. According to Whitaker the driving potential of the liquid water flow is liquid pressure and the effect of thermodiffusion can be neglected.

2.2.3 Description of concrete drying based on diffusion

Bažant and Najjar [10, 11] proposed a simple but accurate model to describe the process of drying of concrete. The model predicts the distribution of relative humidity in the concrete pores based on the Fick's second law. It considers the gradient of relative humidity as the driving potential for the moisture transport and the conductivity is given by a non-linear diffusion coefficient dependent on the temperature and the pore relative humidity of the material. The model does not include the influence of temperature gradients but considers the effect of a temperature change through its influence on the diffusion coefficient and on the relative humidity of the concrete pores (see Eq. 2.3). Assuming constant temperature, the model predicts how diffusion causes the relative humidity of the concrete pores to change with time by solving the following partial differential equation:

$$\frac{\partial h(\mathbf{x},t)}{\partial t} = \mathrm{div}(D(h,T)\cdot \mathrm{grad}h(\mathbf{x},t)) \tag{2.4}$$

where h is the relative humidity of the concrete pores, $\mathbf{x}$ is the vector of position, t is time, D is the diffusion coefficient, and T is the temperature. The diffusion coefficient is calculated as the product of the diffusion coefficient at reference conditions D_1 (i.e. $h = 1.0$ and $T = 20$ °C) and two functions accounting for the influence of the relative humidity $\mathrm{f}(h)$ and the temperature $\mathrm{g}(T)$.

$$D(h,T) = D_1 \cdot \mathrm{f}(h) \cdot \mathrm{g}(T) \tag{2.5}$$

$$\mathrm{f}(h) = \alpha_0 + \frac{1-\alpha_0}{1+\left(\frac{1-h}{1-h_c}\right)^n} \tag{2.6}$$

$$g(T) = \frac{T+273}{T_{\text{ref}}+273} \cdot \exp\left[\left(\frac{Q}{R_D}\right)\left(\frac{1}{T_{\text{ref}}+273} - \frac{1}{T+273}\right)\right] \tag{2.7}$$

In Eqs. 2.6 and 2.7 the relative humidity h is defined between 0 and 1.0 and the temperature T in centigrades. The influence of the relative humidity on the diffusion coefficient $f(h)$ is approximated by a s-shaped function defined by three parameters as illustrated in the left diagram of Fig. 2.10. The parameter h_c defines the location of the inflexion point of the function half-way between the maximum and minimum value of the diffusion coefficient. The parameter α_0 represents the ratio between the minimum and the maximum value of the diffusion coefficient, and n characterizes the spread of the drop in the diffusion coefficient. In Eq. 2.7, illustrated in the right diagram of Fig. 2.10, T_{ref} corresponds to the temperature of reference, i.e. 20 °C. The diffusion coefficient increases exponentially with the influence of temperature depending on the ratio between the activation energy of diffusion Q and the specific gas constant of water vapour R_D.

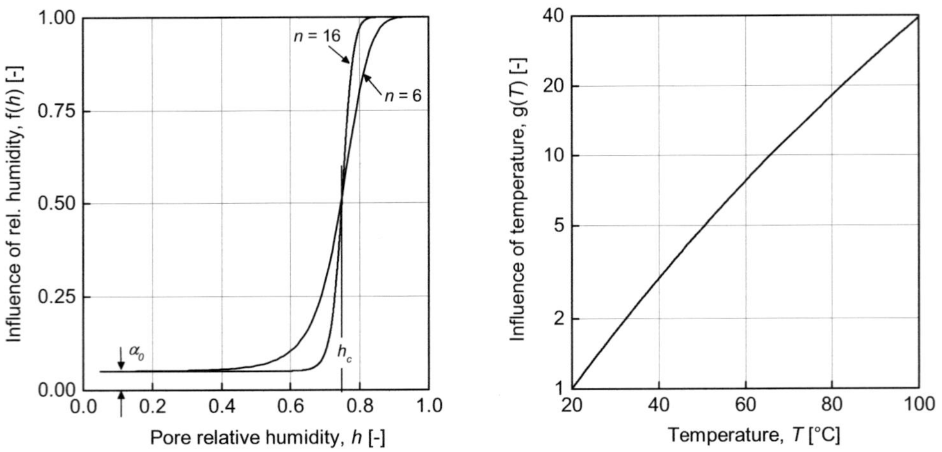

Figure 2.10: Influence of the pore relative humidity and the temperature on the diffusion coefficient of concrete according to Bažant and Najjar [11]

Bažant and Najjar calibrated the model based on data on drying reported in the literature. According to this calibration the parameter α_0, ranges between 0.025 and 0.10, n between 6 and 16, h_c lies around 0.75, and the activation energy of diffusion is roughly equal to 2160 kJ/(kg·K). Despite the limitations of the model, it has been very well accepted by the concrete scientific community which led to its inclusion in the CEB-FIP Model Code 1990 [N14] and it has been further maintained in the new release, *fib* Model Code 2010 [N15].

Further developments based on the model from Bažant and Najjar have been proposed to include the moisture content. Kießl and Gertis [81] calculated the moisture content of the material by translating the results of relative humidity of the model into specific moisture content in volume percent with the help of an assumed sorption isotherm. Xi et. al [154, 155] developed a model to predict the adsorption isotherm of Portland cement paste. They considered the specific moisture content by calculating the moisture capacity as the derivative of the sorption isotherm and integrated it into the moisture transport equation from Bažant and Najjar. These two approaches are however limited to water transport at room temperature because sorption isotherms valid only at 20 °C were employed.

2.3 Influence of temperature on the concrete mechanical properties

Different aspects of the effects of high temperatures in concrete have been widely studied by several authors in the last decades. The most common fields of investigation have been the behaviour of concrete members under fire conditions (high temperatures up to 1350 °C) and the influence of temperature on thick-wall structural members of concrete reactor vessels (temperatures ranging from 20 to 250 °C). At the beginning of the eighties of the last century, Schneider [136] published a summarizing review of the state of the knowledge concerning the behaviour of concrete under the action of high temperatures. The report combines results from researches of reinforced concrete structural components under fire and prestressed concrete pressure vessel for nuclear reactors. It covers a wide range of elevated temperatures neglecting, in most of the cases, the temperatures below 100 °C. Similar reports including high strength concretes were published by the U.S. Nuclear Regulatory Commission in the first decade of the present century [103, 104]. They include the effects of elevated temperatures on the mechanical as well as the physical properties of concrete. Another investigation worth mentioning is the work from Seeberger et al. [138]. They conducted experimental investigations concerning the strength characteristics of concrete subject to temperatures up to 250 °C differentiating between aggregate type and sealing conditions. Even though temperatures under 100 °C are taken into account, only one measure below 100 °C was conducted, namely at 60 °C.

2.3.1 Experimental investigations on concrete submitted to elevated temperatures below 100 °C

The behaviour of concrete under temperatures below 100 °C has not been a common case of study. As mention before, most of the conducted investigations neglect the effects of temperatures below 100 °C on the mechanical properties of concrete focusing on higher ranges of temperature. Nevertheless, Budelmann [35] (see also Rostásy and Budelmann [129, 128]) conducted an extensive investigation covering the combined effects of temperature and humidity on the compressive strength, tensile strength, modulus of elasticity, pore structure, thermal dilation, creep, and shrinkage of concretes subject to temperatures up to 90 °C. This work offers a first clear interpretation of the phenomena occurring in the microstructure of concrete considering the interaction of water and solid matrix at elevated temperatures.

2.3.1.1 Strength and stiffness

Due to the lack of studies specifically oriented at investigating the effect of elevated temperatures up to 100 °C on the mechanical properties of concrete, a compilation of measurements from several investigation was carried out. The majority of the collected data correspond to the initial measurements of large investigations covering wide ranges of temperature. In the following diagrams the influence of temperature is represented by the ratio between the measurements conducted on heated and unheated samples. Fig. 2.11 presents the results of the collected data on compressive strength. The diagram is composed by a total of 216 measurements taken from the experiments conducted by Behnood and Ziari [20], Bingöl and Gül [22], Campbell-Allen and Desai [36], Castillo and Durrani [37], Cheng et al. [38], Furumura et al. [59], Guo and Waldron [69], Hoff et al. [76], Nielsen and Biéanić [109], Phan et al. [110], Phan and Carino [111], Raju and Rao [122], Raju et al. [123], Rostásy and Budelmann [129], Sarhar and Khoury [134], Savva et al. [135], Seeberger et al. [138], Wu et al. [153], Xiao and Falkner [156], Zhang et al. [158], and Zoldners [159].

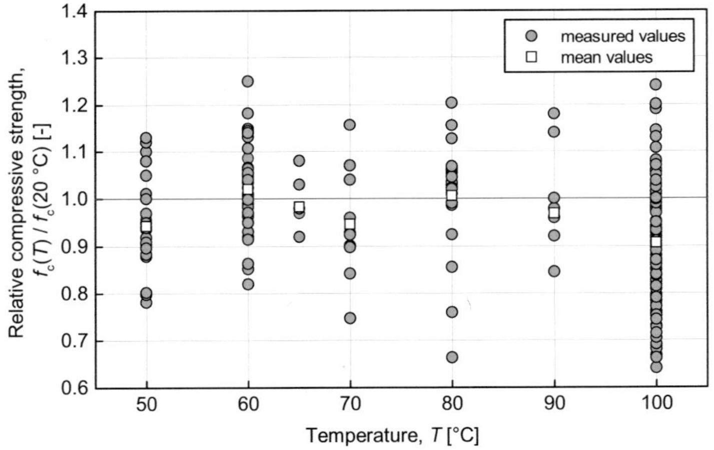

Figure 2.11: Relative compressive strength of samples tested after being subject to elevated temperatures in relation to samples tested at 20 °C

Due to the scattering of the results a clear tendency of the behaviour of the concrete compressive strength at elevated temperatures cannot be seen. The range in which most of the points are located correspond to values between 0.7 and 1.2 while the mean values remain close to 1.0. This means that according to the collected data, subjecting concrete to elevated temperatures below 100 °C can lead to an increase up to 20 % but also to a drop down to 30 % of the compressive strength of the heated samples when compared to unheated samples. The number of samples benefited by an improvement of the compressive strength due to an increase of temperature is closely the same as the number of samples that suffered a reduction of it, which is denoted by the average values mostly located around the line 1.0.

The results of the collected data on tensile strength are presented in Fig. 2.12. The diagram is composed by 95 measurements from the investigations conducted by Campbell-Allen and Desai [36], Guo and Waldron [69], Nielsen and Biéanić [109], Pineaud et al. [113], Rao et al. [124], Rostásy and Budelmann [129], Xiao and Falkner [156], Zhang et al. [158], and Zoldners [159].

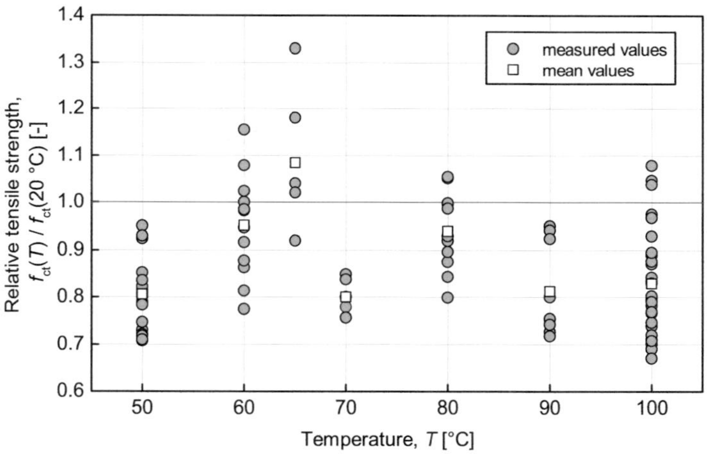

Figure 2.12: Relative tensile strength of samples tested after being subject to elevated temperatures in relation to samples tested at 20 °C

Less investigations on the effect of temperature on the tensile strength have been carried out and therefeore the collected data on tensile strength are smaller. Measuring the tensile strength of concrete can be done in many different ways. The effects of temperature on uniaxial tensile strength, splitting tensile strength, or flexural tensile strength may not necessarily be equivalent, which contributes to the scattering of the results. The points are located between 1.1 and 0.7 and the mean values, though unevenly, are located between 0.9 and 0.8. The results show that a possible enhancement of the tensile strength due to the influence of temperature is less feasible compared to compressive strength. Based on the collected data, a clear tendency of the influence of temperature on the tensile strength cannot be distinguished.

The results of the collected data on modulus of elasticity are presented in Fig. 2.13. The diagram is composed by a total of 86 measurements taken from the experiments conducted by Campbell-Allen and Desai [36], Castillo and Durrani [37], Cheng et al. [38], Furumura et al. [59], Guo and Waldron [69], Phan et al. [110], Phan and Carino [111], Rostásy and Budelmann [129], Wu et al. [153], and Zhang et al. [158].

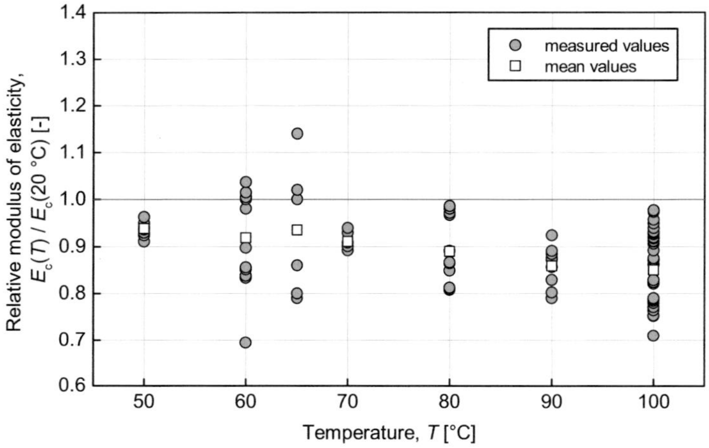

Figure 2.13: Relative modulus of elasticity of samples tested after being subject to elevated temperatures in relation to samples tested at 20 °C

Even though the influence of temperature on the modulus of elasticity of concrete has been seldom investigated, the collected results show smaller scattering and the mean values depict a tendency of reduction of the modulus of elasticity with the increment of the temperature. Almost all the points are located below the 1.0 line, which denotes that an enhancement of the modulus of elasticity due to the effect of temperature is not likely to occur. According to the collected data, in average the modulus of elasticity decreases linearly with increasing temperature at a rate of 2 % every 10 K.

Combining results from different investigations hampers the possibility of understanding the physical phenomena behind the system because of the scattering of the results. This scattering is most likely given by the differences between test procedures and storage conditions, compositions, as well as geometries of the tested samples. Moreover, the scattering seen in Figs. 2.11 to 2.13 can be reduced considerably if the comparison is done based on a specified moisture content because, as it will be shown in Chapter 5, the effects of temperature on the mechanical properties of concrete highly depend on the moisture content of the concrete member. The purpose of presenting these results here is on one hand, to identify the ranges of values in which the influences may be delimited and on the other hand, to emphasize the fact that modelling the influence of temperature on the strength and stiffness of concrete cannot be accurately addressed based on the available investigations.

Influence of moisture content on the strength of heated samples

In the following paragraphs some selected results from Budelmann [35] are presented and discussed with the intention of throwing light on the problem and help with the understanding of the scattering seen in Figs. 2.11 to 2.13. As mentioned before, the experiments conducted by Budelmann accounted for the interaction between the effects of temperature and moisture content. Understanding this interaction helps to guide the design of a testing program that can deliver results consistent enough to achieve the objectives of the investigation.

After keeping the concrete samples in sealed conditions for 90 days, Budelmann divided the samples in three groups according to the humidity conditions of the environment, namely: 65 % RH, 95 % RH and under water. The samples were kept initially at 20 °C during 60 days and subsequently they were subject to elevated temperatures of 50, 70 and 90 °C without modifying the humidity conditions of the environment up to 120 days longer. Reference values of the concrete properties were obtained from samples after the conditioning by 20 °C was finished. Fig. 2.14 presents results of compressive strength of samples from the three groups subject to 50 and 90 °C and tested after 20, 60 and 120 days of exposure to elevated temperature.

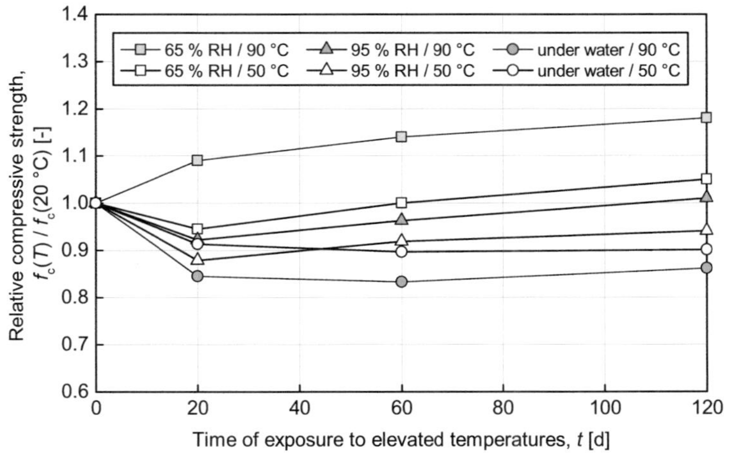

Figure 2.14: Development of the relative compressive strength with time for concretes heated at different stationary ambient conditions [35]

The samples heated under water present a reduction of strength without showing any strength recovery with time. The reduction of strength is the higher the higher the temperature of exposure. The samples heated under constant relative humidities of of 65 and 95 % showed a recovery of the strength with time. This recovery effect was more notable on the samples subject to higher temperatures. The samples stored at 65 % RH and heated at 90 °C showed an enhancement of the compressive strength.

According to the results from Budelmann, the influence of moisture content before and during the exposure of the samples to elevated temperature influences the strength of concrete in such a way that variations between a decrease of 20 % and an increase of 20 % on the concrete compressive strength can be seen in samples of the same concrete subject to the same temperature. These results evidence that the storage conditions of the samples before and during test can be to a large extent responsible for the scattering seen in Figs. 2.11 to 2.13.

Budelmann explained the observed changes in the concrete strength due to elevated temperatures as the result of the combined effects from several factors as follows:

- The effects of strength recovery and gain are due to dewatering of the samples, activation of the cement hydration, and formation of new CSH-phases because of the hydrothermal reactions of $Ca(OH)_2$ in the cement paste with the SiO_2 of the fine siliceous aggregates. All these processes are time dependent which explains that the compressive strength increased with the time of exposure.
- The strength decreases due to microcracking caused by thermal and hygral incompatibilities. The thermal incompatibility refers to differences between the thermal expansion coefficients of the cement paste and the aggregates and occurs only during heating which means that this damage is not dependent on the duration of exposure. The hygral incompatibility refers to differences in the moisture content of the concrete samples within its geometry causing internal stresses, which are presented during the whole drying process. Therefore, the damaged caused by the hygral incompatibility is time dependent.

Based on the observations from Budelmann, it can be stated that the influence of temperature on the concrete strength and stiffness is linked to the moisture content of the concrete. Three time dependent factors influence the mechanical properties positively, namely: dewatering, acceleration of hydration and hydrothermal reactions. Two factors influence the properties negatively, the time independent thermal incompatibility and the time dependent hygral incompatibility. The combination of these five factors is, according to Budelmann, responsible for the changes on the strength and stiffness of concrete subject to elevated temperatures below 100 °C. Depending on the mechanical property being evaluated, the influences of each factor can vary according to the sensibility of the mechanical property to the factor. For instance, tensile strength is very sensitive to microcracking and therefore, after heating a concrete member, the damage caused by microcraking is more meaningful for the tensile strength than for the compressive strength or the modulus of elasticity.

2.3.1.2 Creep and shrinkage

Concrete belongs to the group of materials which show pronounced creep and shrinkage deformation under normal conditions of use. Creep and shrinkage are the two forms of time-dependent deformation exhibited by concrete. Shrinkage is the portion of time-dependent deformation due to physical or chemical causes other than applied load, and creep results from subtracting the sum of the elastic and shrinkage deformations from the total deformation under load. These deformation processes in concrete have been intensively studied since the middle of the 20th century, in such a manner that nowadays, the load and time-dependent behaviour of concrete can be predicted with reasonable certainty, in particular, at a constant ambient humidity and temperature and constant load. General reviews on creep and shrinkage of concrete can be found, for instance, in works published by Bažant and Wittmann [13], Müller [96], or Neville et al. [107].

Considering only the deformation behaviour under constant conditions, the numerous systematic investigations conducted on concrete samples summarized as follows:

- The creep and shrinkage deformations interact closely with each other and have both reversible and irreversible portions.
- The time-dependent deformations grow with increasing cement paste volume [105, 114, 150] and decreasing stiffness of the aggregates [42, 125].
- With decreasing porosity of the hardened cement paste, i.e. with decreasing w/c-ratio and increasing hydration (concrete age), the time-dependent deformations decline [102, 108, 150].
- With increasing water content of the concrete and increasing rate of water loss during loading, the time-dependent deformations increase [95, 150]. Creep and shrinkage are therefore the higher, the lower the relative humidity of the environment and the smaller the concrete member [72].

Several hypotheses can be found in the international literature about the causes of creep of concrete. In general, it is agreed that creep is due to a load-dependent sliding process in nanoscale water films between or within the hardened cement gel particles formed by calcium silicate hydrates (CSH) and shifts in the thermodynamical balance of the concrete microstructure (see e.g. Feldman [51], Powers [119], Ruetz [133] or Setzer [142]). A condition for the occurrence of creep deformation is the presence of water in the pore system of the cement paste. This was evidenced by the work from Brown and Hope [30], in which oven-dried mortar samples exhibited virtually no creep deformation.

Regarding the influence of temperatures up to 100 °C on the creep and shrinkage deformations a considerable number of works have been published. Most of these investigations were conducted during the sixties and seventies of the last century due to the need for more detailed knowledge about the strain behaviour of concrete at elevated temperatures impulsed by the development of prestressed concrete pressure vessels for the containment of nuclear reactors. Initially the investigations were limited to sealed concretes under uniaxial and multiaxial stress conditions, however, with the development of new test and measuring devices, tests on unsealed samples, loaded before and after reaching the desired test temperature, were also carried out. The investigations compare the creep and shrinkage behaviour of the concrete samples at elevated temperature with samples tested at ambient temperature. A brief review of some selected investigations regarding the influence of temperature on the time-dependent deformations of concrete is presented below. The most relevant observations from these investigations are presented in chronological order to show the evolution of the studies which compile the actual state of knowledge.

- Browne [31, 32] conducted creep investigations on sealed samples at temperatures up to 95 °C. The samples were heated-up one day before loading. Browne found that temperature has a large effect on creep rate, particularly in the first 3 days of loading. After the first 3 days, the creep rate of samples tested at elevated temperature becomes similar to that of samples tested at 20 °C.

- Hannant [70] tested creep on samples under uniaxial and multiaxial compressive stress at temperatures between 27 and 95 °C. Sealed and unsealed samples were loaded 24 hours after the start of heating. Regarding the sealed samples, it was found by Hannant that the relation between creep and temperature is approximately linear between 27 and 80 °C. The rate of increase of creep accelerates as the temperature approaches 100 °C which looses the linearity seen by temperatures up to 80 °C. The results from the unsealed samples were discarded because allowing the samples to stabilize at the test temperature 24 hours before loading caused considerable changes in the moisture condition of the samples making the results very dependent on the specimen size and hence useless to make comparisons with the sealed samples. Samples previously dried reached creep values that were only a small fraction of the values shown by wet samples, even at temperature of 70 °C.

- York et al. [157] measured compressive and tensile creep deformations on samples under uniaxial and multiaxial loading at temperatures of 25 and 65 °C. The samples were cured in sealed and unsealed conditions. After the curing process, the samples were sealed and heated-up a week before loading. They found that both compressive and tensile creep were larger for: samples tested at 65 °C, samples cured in unsealed conditions, and samples tested under uniaxial state of stress.

- Seki and Kawasumi [139] conducted tests on sealed and unsealed samples at temperatures between 20 and 70 °C. The loading took place after the samples were heated-up and no humidity control was employed during the tests of the unsealed samples. They found that both the instantaneous elastic strain and the creep strain increase with increasing temperature. They measured larger creep strains in unsealed samples when compared to sealed samples. Sealed samples tested at 70 °C showed creep deformations 1.6 to 3.2 times larger than the ones tested at 20 °C. On unsealed concretes this quotient was of around 1.4 to 2.1.

- Kordina and Budelmann [88] published an extensive literature review on the influences of temperature up to 100 °C on the creep of concrete. Based on this review, Budelmann identified the need of conducting systematic experiments to appraise the combined effects of temperature and moisture content on the time-dependent deformations of concrete as well as the differences on the creep deformations resulting from loading before and after heating-up the samples. After developing an appropriate testing device [130, 132], Budelmann [35] conducted creep tests on concretes subject to temperatures up to 90 °C. The concrete samples were loaded before and after being heated and the relative humidity during the creep test was controlled and kept constant either at 65 % or 95 % RH. Budelmann found that the samples tested at elevated temperatures and low environmental relative humidity exhibited lower rate of creep after a load duration of about 20 days than the samples tested at the same temperatures and higher relative humidity. Samples loaded at first and then heated-up showed considerably higher creep deformations than samples loaded after the temperature of testing was reached. He concluded that the sequence of heating and loading plays an important role on the development of the creep deformations specially during the first 1 to 2 days of loading.

- Schwesinger et al. [137] measured creep deformations on sealed and unsealed samples heated-up after the initiation of loading at temperatures up to 130 °C. They found that at temperatures higher than 60 °C the creep strains measured on sealed samples were higher than those measured on unsealed samples. Compared to normal temperature, the unsealed specimens tested at elevated temperatures showed creep values 1 to 4 times higher after 100 days of loading. In contrast, the creep values of the sealed samples at elevated temperatures exceeded the values at 20 °C by factors between 6 to 8. They also pointed out that these factors decrease with time and therefore it could be expected that for longer periods of time the difference between heated and unheated samples will vanish.

- Küttner and Ehlert [91] conducted tests oriented on understanding the differences between the creep deformations of samples loaded before and after stablishing the testing temperature. They defined transitional creep as the part that results from the influence of elevated temperature when the sample is loaded and then heated. It can be calculated by subtracting the creep deformations measured in a sample heated and then loaded from a sample loaded and then heated. They found that the transitional creep deformation is dependent on the temperature and time and it describes a finite process which finishes almost completely after about two days.

There are various interactions between the factors influencing creep and shrinkage. Even though these interactions are well known, they have not yet been sufficiently quantified through systematic investigations. Unresolved questions can be seen in particular in the field of microstructural mechanisms of creep and the influence of changes in the applied load and relative humidity of the environment.

2.3.2 Theoretical approaches towards the conception of an experimental program

In the following subsections some considerations regarding the influence of temperature on the mechanical properties of concrete are presented with the objective of settling the basis for the design of the experimental investigations. These considerations are based on the general state of knowledge elaborated in the present chapter and more specially on the observations shared by Budelmann [35]. The investigations conducted by Budelmann took into account the combined influences of temperature and moisture content on the mechanical properties of concrete allowing to identify the factors that influence the strength and stiffness of concrete and emphasizing on the importance of water content on the development of the time-dependent deformations of concrete.

2.3.2.1 Strength and stiffness

Between the factors influencing the strength and stiffness of concrete subject to elevated temperatures as described by Budelmann [35], namely dewatering, hydrothermal reactions, acceleration of hydration, hygral incompatibility and thermal incompatibility, interactions probably exist. However, as an initial approach and for sake of simplicity, it can be assumed that these five factors influence the concrete mechanical properties independently and consequently the total effect that they exert on the mechanical properties of concrete obeys the principle of superposition. Assuming superposition implies that the effects of temperature and moisture content on the strength and stiffness of concrete can be described as the summation of the effects caused by each independent factor. The influences of these factors on the strength and stiffness of concrete develop differently in time according to the conditions of humidity and age previous to the heating phase.

A schematic description of the time development of a given mechanical property β due to the influence of a sustained elevated temperature on a concrete member is presented in Fig. 2.15. The dark curve represents the development of the mechanical property calculated as the summation of the positive effects from acceleration of hydration, hydrothermal reactions and dewatering with the negative effects from hygral and thermal incompatibilities. Due to the thermal incompatibility being time independent, the concrete member suffers a reduction of the mechanical property immediately after the temperature is increased. Later on, the recovery caused by the positive effects of dewatering, hydrothermal reactions and hydration development with time allowing the concrete member to reach values in some cases even higher than the reference at 20 °C.

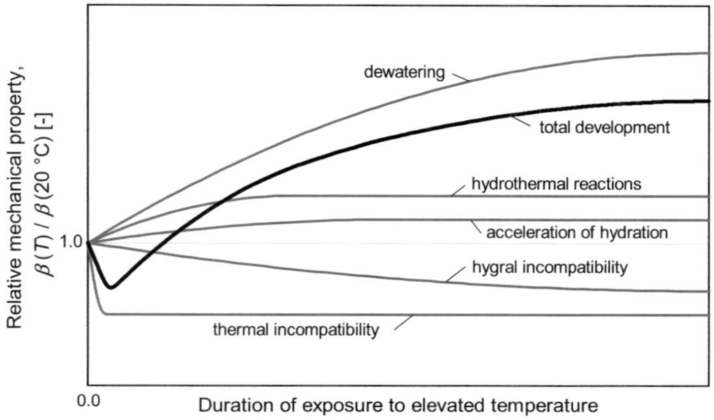

Figure 2.15: Schematic description of the time-development of the influence of temperature on the strength and stiffness of concrete

For the design of an experimental program capable of obtaining enough information to formulate models that can predict the behaviour of concrete at elevated temperatures the following simplifications and assumptions were made.

- As the damaged caused by thermal incompatibility occurs only during heating, samples tested after reaching the elevated temperature can be used to quantify the influence of the thermal incompatibility.
- The positive effect of dewatering and the negative effect of hygral incompatibility on a given concrete mechanical property obey the same time dependency and cannot be appraised separately. These two factors can be combined and considered in a single factor which will be referred to as drying.
- The factor accounting for the acceleration of hydration processes can be neglected if the samples to be tested are old enough to assume that none or very few hydration reactions can still take place.
- As initial assumption, the effects or hydrothermal reactions on the strength and stiffness of concrete will be neglected. According to Budelmann, these reactions did not show any significant influence on the concrete strength.

Based on the assumptions listed above, the experimental program shall be focused on investigating the effects of thermal incompatibility and drying on the strength and stiffness of concretes subject to elevated temperatures. Being drying the only time dependent factor to be evaluated, measuring it becomes highly relevant because the time development of drying can be correlated with the behaviour of the concrete mechanical properties over time.

2.3.2.2 Creep and shrinkage

Changes in the moisture content during loading and moisture content previous to loading are probably the most important factors influencing the deformation behaviour of concrete under sustained loads. Engineers on charge of developing code-type models to predict the creep and shrinkage behaviours of concrete have struggled with the issue of incorporating these two factors into the models without losing simplicity and generality. A simplified but well accepted way to incorporate the first factor has been the partition of shrinkage and creep into their basic and drying components. This provides the possibility of taking into account the changes in the moisture content within the drying components of creep and shrinkage by defining adequate time-development functions [97]. Incorporating the content of moisture at the beginning of loading is, however, more complicated because it involves considering the drying history of the concrete member before loading. The models however can partly consider it, if not directly at least hidden, in the factors accounting for the concrete age.

Although the existing investigations, briefly summarized before, already covered the most important aspects and factors influencing the time-dependent deformations of concrete subject to elevated temperatures, modelling creep and shrinkage at elevated temperatures is still a challenging issue. The main difficulty resides in the fact that the current models are conceived for a reference temperature of 20 °C and the influences of moisture content and moisture gradients are not directly embraced. The gain on importance on the deformation behaviour of concrete that the water transport processes undergo due to the effect of temperature can only be captured by a model capable of directly consider the effects of the moisture content and moisture gradients. Therefore, an experimental program aiming at the formulation of creep and shrinkage models shall include the following considerations:

- Tests conducted at 20 °C may be used as reference.
- The influence of moisture content shall be identifiable. For this purpose samples will be conditioned at 20 °C and different relative humidities for a long time before testing. Consequently, the samples will be tested at the same relative humidity in order to allow for comparability between the different conditioning scenarios. This applies also for the tests conducted at elevated temperatures.
- The influence of temperature may be appraised by testing at two different temperature levels besides the reference at 20 °C.

- The water content and drying process of the samples shall be recorded during the conditioning of the samples previous to the tests, as well as during the test phase.
- To limit the number of tests only one regime of heating and loading may be considered. Heating before loading would be the best option because this way the influence of thermal expansion can be neglected.

Not considering the effect that loading before heating causes in the creep deformations of concrete subject to elevated temperatures limits beforehand the extent of the models to be formulated, however, as it has been described by Budelmann or Kuettner and Ehlert [35, 91] the additional deformations caused by this heating regime are only relevant in the first 2 days of loading.

2.4 Summary

The chapter makes a journey through the importance of water content on the concrete physical and mechanical properties. It begins pointing out its relevance on the mix design, not only to ensure workability but also to define the concrete microstructure as prime factor controlling the porosity of the hardened cement paste. Subsequently, the chapter introduces three well known cement paste models, which agree on the fact that water strongly interacts with the solid matter within the microstructure of the hardened cement paste and its presence is important to understand the behaviour of the mechanical properties of concrete. After having a conception of the concrete microstructure, it becomes relevant to assess the storage of water within it. For these purposes, the mechanisms that regulate the ingress of moisture in the concrete microstructure are introduced and the roles of these mechanisms are then explained on the basis of the sorption isotherms of the material. The concept of sorption isotherm is also used to describe the influences of temperature on the water storage capacity of concrete and as source of redistribution of the water molecules between the pores of the microstructure. Besides the physical changes on the microstructure originated by water movement and the action of temperature, chemical reactions can also take place. The chapter refers briefly to the hydrothermal reactions that the concrete microstructure can experience when it is subject to elevated temperature in the presence of water.

Calculating the movement of moisture within the concrete microstructure provides a fundamental tool to develop predictions of the influence of moisture content on the mechanical properties. The chapter summarizes the mechanisms associated with the transport of fluids through porous materials and presents a simple model to describe the drying process of concrete members. Subsequently, it comprises a selection of experimental investigations dealing with the influence of temperature on the mechanical properties of concrete. The fact that the vast majority of these investigations neglect or forget about the influence of water content on the results leads to the necessity of conducting new experimental investigations. In the last part of the chapter assumptions and simplifications taken into account for the design of an experimental program are presented differentiating between tests to appraise strength and stiffness and test to assess the time-dependent deformations of concrete.

Chapter 3

Experimental program

This chapter presents the conception of an experimental program with the purpose of acquiring knowledge and quantitative correlations that can further be used for the formulation of models to predict the behaviour of concrete at different conditions of temperature and moisture. The entire process followed to conduct the experimental investigation is described including references to the standards used and assumptions taken which enclose the background of the experiments. It includes the design of the concrete mixtures, production and storage of the concrete samples and the experimental procedure specifying the tests conducted. In order to carry out the proposed experiments, special testing technology was developed. Brief descriptions of the testing technology are mentioned is in this chapter and further details are included in Appendix A.

3.1 Concrete samples

3.1.1 Source Materials

The concrete samples were produced with Portland cement CEM I 32.5 R as binder, superplasticizer as additive, regional available aggregates (Rhine sand and Rhine gravel), and tap water. Some characteristics of the source materials are listed in Table 3.1.

Table 3.1: Properties of the materials used for the production of the concrete samples

Source material	Specification	Density [kg/m³]	Solids content [M.-%]	Water content [M.-%]
Binder	Portland cement CEM I 32.5 R	3,10	-	-
Aggregates	Rhine sand 0/2	2,63	-	0,15
	Rhine gravel 2/8			1,90
	Rhine gravel 8/16			1,50
Additives	Polycarboxylate superplasticizer	1,06	30,0	-

The use of additives was necessary in order to set a consistency of the fresh concrete that could ensured appropriate workability at low values of w/c-ratio. Using only Portland cement and siliceous aggregates as source materials limits the compass of the investigation. However, due to the complexity of the investigated phenomena, this restriction was necessary in order to conduct the research with a manageable amount of test samples. The formulations proposed in Chapter 5 are in principle limited to concretes made with the source materials employed for this investigation. Nevertheless, the applicability and possible reach of the material model are extensively discussed in Chapter 6.2.3.

3.1.2 Mixtures

Three mixtures were developed to produce concretes for the investigation with the objective of covering the range of concrete strengths that is most frequently used in the practice. In order to formulate models that can predict the effects of temperature and moisture content, the investigated concretes need to be comparable with each other, particularly with respect to their deformation behaviour and properties of moisture transport. On this behalf, the following premises were adopted:

- The particle size distribution of the aggregates was kept identical for all concretes aiming at reaching good workability and packing density. In this way the tortuosities of the concretes do not differ substantially. As seen in Fig. 3.1 the particle size distribution of the aggregates is oriented to the standardized grading curves A16 and B16 according to DIN 1045-2 [N1]. Setting the maximal particle size to 16 mm allows to use relative small diameters for the samples (down to 64 mm according to DAfStb[1] booklet 422 [N13]), which is necessary, given that the time required for a concrete member to dry is subjected to its geometry.
- The cement paste content was fixed to 300 dm^3 per m^3 of concrete. Consequently, the influences of the cement paste content on the concrete stiffness and time-dependent deformations are equivalent between the concretes.
- The differences between the concretes lie in the added proportion of water and cement. The w/c-ratio was varied in 3 levels: 0.4, 0.5 and 0.6. According to the model of cement paste from Powers and Brownyard [120], by a w/c-ratio of 0.4 all the cement particles are able to react with water leaving very few capillary porosity at the end of the hydration reaction. The w/c-ratios of 0.5 and 0.6 lead to excess of water and capillary porosities corresponding to approximately 20 % and 30 % of the total cement paste volume respectively (see Fig. 2.1 in Chapter 2). Thus, the concretes differ only in the microstructure of their hardened cement pastes, which allows them to have different strengths, stiffness, densities, and permeabilities and still be correlative.

[1] German Committee for Structural Concrete

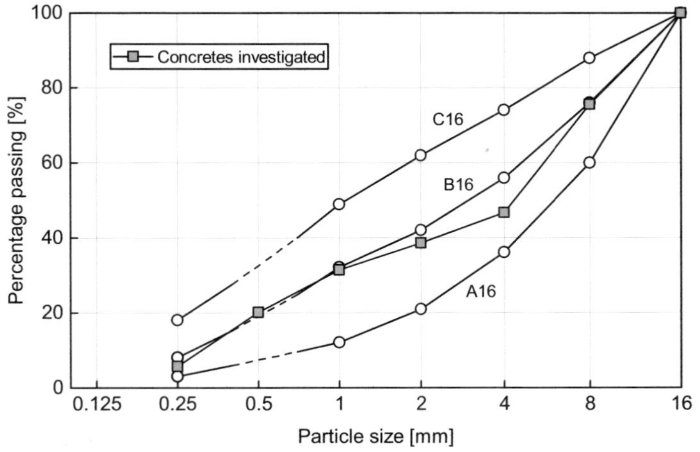

Figure 3.1: Particle size distribution of the aggregates from the investigated concretes and standard grading curves according to DIN 1045-2 [N1]

The concrete mixtures were named according to the w/c-ratio as follows: MRC (mixture of reference concrete) for the w/c-ratio of 0.5; MHC (mixture with high water content) corresponding to a w/c-ratio of 0.6 and MLC (mixture with low water content) having a w/c-ratio of 0.4. Table 3.2 presents the composition of the three concretes used for the experimental investigation.

Table 3.2: Composition of the concrete mixtures

Component		MRC	MLC	MHC
Water/cement ratio	[-]	0,5	0,4	0,6
Cement paste content	[dm³/m³]	300	300	300
Cement	[kg/m³]	340	386	304
Water	[kg/m³]	165	149	179
Rhine sand 0/2	[kg/m³]	718	719	718
Rhine sand 2/8	[kg/m³]	646	647	646
Rhine gravel 8/16	[kg/m³]	479	480	479
Superplasticizer	[M.-%] of cement	0,5	0,75	0,0
Mass proportions (c : a : w)	[-]	1 : 5.4 : 0.5	1 : 4.8 : 0.4	1 : 6.1 : 0.6

To characterize the fresh concrete properties of the mixtures slump, density and air content were determined according to DIN EN 12350-5 [N4], DIN EN 12350-6 [N5] and DIN EN 12350-7 [N6], respectively. The compressive strengths at ages of 7 and 28 days were determined on cube samples with an edge length of 150 mm according to DIN 1048-5 [N2]. The fresh concrete characteristics and the cube compressive strengths of the hardened concretes are given in Table 3.3.

Table 3.3: Fresh and hardened concrete properties of the mixtures

Characteristic		MRC	MLC	MHC
Flow diameter	[cm]	38	36	33
Air content	[%]	2.1	2.3	2.4
Density	[kg/dm³]	2.36	2.36	2.33
Cube compressive strength after 7 days	[N/mm²]	46.3	58.4	38.6
Cube compressive strength after 28 days	[N/mm²]	54.8	77.8	47.9

The concretes in fresh state showed flow spreads from 33 to 38 cm (rigid to plastic consistency according to DIN 1045-2 [N1]). The air void contents varied from 2.1 to 2.4 % and the densities from 2.33 to 2.36 kg/dm^3. The compressive strengths measured at an age of 28 days correspond to concretes ranging from normal strength C30/37 to high strength C55/67 according to DIN EN 206-1 [N9].

3.1.3 Production scheme

The concretes were produced in the laboratory in accordance with the mixing and processing regime described in Table 3.4.

Table 3.4: Procedures followed in the production of the concretes

Production step		Procedure	Time
1st	Addition of aggregates	Mixing	45 s
2nd	Addition of cement	Mixing	45 s
3rd	Adittion of water	Mixing	90 s
4th	Addition of additive	Mixing	60 s
5th	Casting of concrete in wall formworks	Compacting	
6th	Casting of concrete in cylindric molds	Compacting	2 x 10 s
7th	Wrapping of the concrete elements with jute sacking	Curing	90 d

The preparation of the concretes was performed using a pan mixer with a nominal capacity of 250 dm^3. After mixing the aggregates, cement was added followed by water and finally the additive. The concretes were gradually placed in layers in formworks and molds and compacted with hydraulically driven vibrators. Concrete walls were cast using 1200 mm x 1000 mm x 200 mm formworks and concrete cylinders by means of standardized molds (Diameter/Height = 150/300 mm).

During 90 days following the production of the concrete elements, the concrete walls and cylinders were stored at 20 °C, covered with jute sacks and kept wet to avoid drying of the edges. This storage condition allowed the concrete elements to reach high levels of hydration before the production of the samples could continue. In total, three concrete walls and 30 concrete cylinders were made with the mixture MRC and one wall and 15 cylinders were made with each one of the mixtures MLC and MHC. After curing, up to 99 cylindrical cores with a diameter of 75 mm were drilled out of each concrete wall and 1 cylindrical core with the same diameter was taken from each cylinder. Then, sections from the top and bottom of each core were cut off to set a desired sample height of 150 mm, and subsequently, the top and bottom faces of the specimens were sanded to ensure parallel alignment between the surfaces. Based on these cylindrical specimens (D/L = 75/150 mm), the concrete samples were prepared according to the geometries illustrated in Fig. 3.2. The samples were classified in 4 types meeting different purposes as follows:

- Type 1: For the studies of compressive strength, modulus of elasticity and deformation behaviour of the concretes, samples with a diameter of 75 mm and a height of 150 mm were used.
- Type 2: Tensile strength of the concrete was measured in samples with a 10 mm deep notch. These samples were notched because of two reasons: in first place, reducing the effective area of the samples facilitates the execution of the tensile strength tests as it ensures the failure of the concrete samples in the notched section and not in the vicinity of the top and bottom surfaces glued to the testing machine, which according to DAfStb booklet 422 [N13], would invalidate the measured value. In second place, as reported by Bonzel and Kadleček [26], micro cracks formation due to drying influences the results of the tensile strength tests significantly. This influence can be attenuated if the moisture gradient within the notch section is reduced, which was accomplished by sealing the notch (see Fig. 3.2).
- Type 3: The development in time of the relative humidity and temperature within the concrete pores was registered in the samples taken from the cylindrical molds by means of relative humidity sensors. In order to place the sensors in the desired positions within the samples, a positioning device was developed to fix 4 stainless steel tubes in the cylindrical molds. During casting, rounded steel bars were introduced through the steel tubes and pushed 10 mm into the concrete to prevent the cement paste from entering the tubes. Afterwards, the bars were removed leaving

an uncovered part of the hole and the steel tubes empty, which allowed the further ingress of the sensors (see Sample type 3 in Fig. 3.2). With the help of this positioning device, one steel tube was placed in the central axis of the sample to a depth of 75 mm and three more were placed 10, 20 and 30 mm separated from the central axis in a 3 x 120° arrangement (see also Fig. 3.6). The depths of the latter were set to 30 mm to avoid any interference on the measurements between the holes.

- Type 4: The water content of the concretes was periodically measured by weighting small segments with a height of approximately 35 mm. At the end of the test phase, all these samples were dried at 105 °C and the dry weights of reference of the concrete mixtures were determined.

Finally, the upper and bottom faces of all samples were sealed with aluminium cored polypropylene films forcing the moisture exchange between the cylindrical samples and the environment to take place only through the lateral surfaces. In this way, the moisture distribution within a given sample can be assumed to be equal in every position along the axis of symmetry.

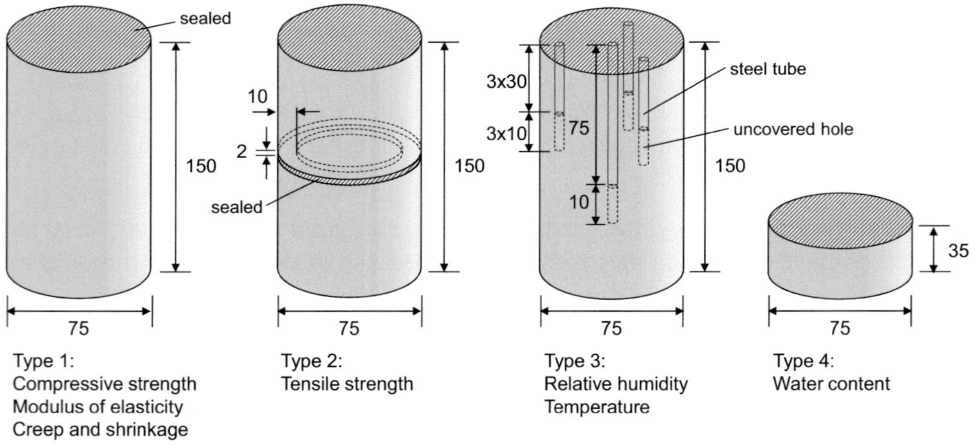

Figure 3.2: Geometry of the prepared samples (dimensions in mm)

After preparation, the samples were stored above water in sealed tubs at 20 °C, which guaranties an environment with relative humidity close to 100 % and constitutes a standardized alternative for storing concrete samples according to DIN EN 12390-2 [N7]. The samples were kept in the tubs for over 150 days before the conditioning by 20 °C started (see Chapter 3.2.1). A time scheme of the samples production and further storage is presented in Fig. 3.3.

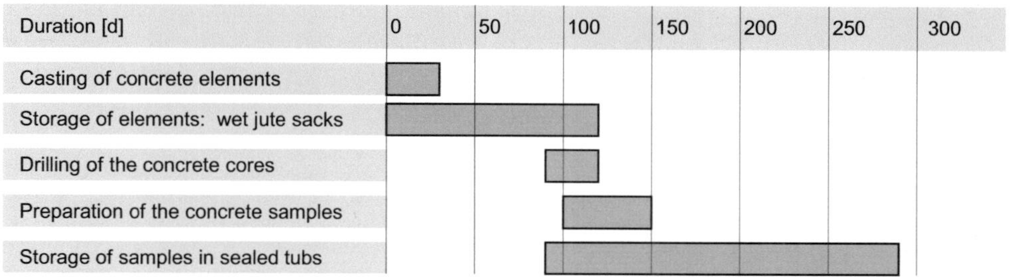

Figure 3.3: Sequence of production of the samples

The largest number of laboratory investigations were carried out with the reference concrete MRC. The amount of samples used for the investigation is presented in Table 3.5 according to the tests conducted and sample type. The total number of samples was 622.

Table 3.5: Number of samples used for the investigation

Sample type	Test purpose	Number of samples of the concrete mixture		
		MRC	MLC	MHC
1	Compressive strength / Modulus of elasticity	140	44	44
	Creep	16	4	4
	Shrinkage	16	4	4
2	Tensile strength	98	40	40
3	Relative humidity	30	15	15
4	Moisture content	60	24	24

The tests conducted with the samples of the concretes MLC and MHC served to address the influence of w/c-ratio (i.e. microstructure and strength) on the variation of the mechanical properties of concrete that were investigated.

3.2 Experimental procedure

Assuming that the principle of superposition applies for the influences of temperature and moisture content on the mechanical properties of concrete (see Chapter 2.3), the following experimental approach was defined:

- Reference samples were tested at a temperature of 20 °C and a relative humidity of 100 %. These experiments configure the basis of reference of the investigation.

- Samples previously dried at different ambient humidity conditions and reference temperature (20 °C) were tested without modifying the temperature. The results of these experiments comprise the influence of moisture content alone.
- Samples were first dried at different ambient humidity conditions and temperature of reference. Subsequently, they were heated under constant environmental humidities and tested at three points in time. Due to the drying of the samples during heating, a new and lower moisture content was reached for each time at which the samples were tested. The results of these tests include the combined effects of temperature and drying.

Based on the approach described above, an experimental program was developed. The planned experiments were oriented to determine the influence of temperature on compressive and tensile strength, modulus of elasticity, creep and shrinkage of concrete for temperatures between 20 and 100 °C in relation to the moisture content of the material. Fig. 3.4 provides an overview of the conditioning of the samples previous to the tests and during testing.

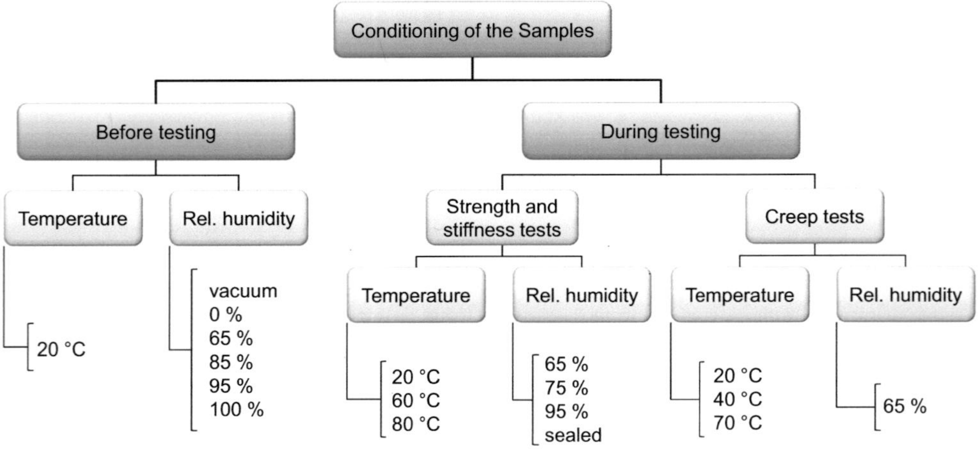

Figure 3.4: Ambient conditions of temperature and relative humidity imposed in the investigation

The temperature range between 20 and 100 °C was covered with 3 different testing temperatures: 20, 60 and 80° C for the strength and stiffness tests and 20, 40 and 70° C for the creep tests. It is assumed that the results of this investigation may be extrapolated for temperatures below 100 °C right before the temperature of ebullition of water is reached and therefore no abrupt changes of state occur.

The moisture content was varied by drying the samples previous to the test through conditioning at reference temperature and 5 different ambient relative humidities: 100 %, 95 %, 85 %, 65 %, 0 %, and under vacuum[2]. During testing strength and stiffness, the samples were either sealed or subject to ambient relative humidities of 95 %, 75 % or 65 %. For the creep tests the relative humidity of the environment was set solely to 65 %.

3.2.1 Conditioning of the samples

After the production and further storage at reference conditions of 20 °C and 100 % relative humidity (RH), the first samples having ages between 260 and 290 days were tested. The results of the tests conducted on this first fraction of samples constituted the reference of the investigation. The rest of the samples were tested after being conditioned at several environmental relative humidities and 20 °C and then subject to different climatic environments of temperature and humidity. For conditioning the concrete samples at 20 °C and constant ambient humidity, 4 climate boxes were used. These boxes were stored in an acclimatized room at 20 °C and 65 % RH. For conditioning at relative humidities other than 65 %, the relative humidities of the boxes were adjusted by means of saturated salt solutions according to DIN EN ISO 12571 [N11]. At higher temperatures the samples were conditioned using climate chambers. A more detailed description of the climate boxes and chambers can be seen in Appendix A.1. The conditioning of the samples followed the time schedule presented in Fig. 3.5.

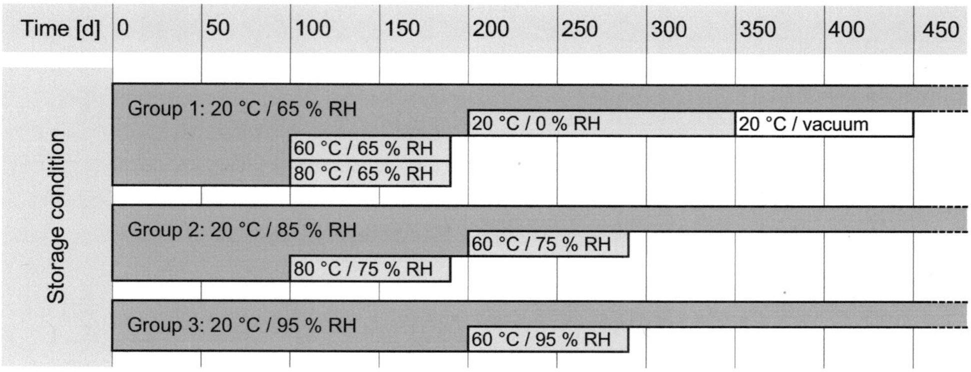

Figure 3.5: Sequence of conditioning of the samples

The samples were distributed in three groups as follows:

- Group 1: Samples were stored at 20 °C and 65 % RH. After 100 days two portions of these samples were moved to be stored at 60 °C / 65 % RH and 80 °C / 65 % RH up to 90 days long. Another portion was moved after 200 days to a dryer environment at 20 °C / 0 % RH and kept there for 140 days. Finally, from this latter

[2] Samples were subject to vacuum in order to dry them as much as possible without deviating the temperature from the reference of 20 °C

portion, some samples were subject to vacuum 110 days long at 20 °C. The rest of the samples were maintained at 20 °C / 65 % RH for over 450 days.

- Group 2: Beginning with a storage condition of 20 °C and 85 % RH, one portion of the samples was moved after 100 days to be heated at 80 °C / 75 % RH for 90 days. After 200 days, a second portion of the samples was subject to 60 °C / 75 % RH, also during 90 days. The remaining samples were kept beyond 450 days by the original storage condition of 20 °C / 85 % RH.
- Group 3: After being stored 200 days at 20 °C / 95 % RH, a part of the samples was subject to 60 °C / 95 % RH for 90 days. The rest of the samples were maintained at 20 °C / 95 % RH for over 450 days.

For those groups of samples that were subject to elevated temperatures the temperature was increased with a constant rate of 5 K/h which ensures an approximately constant temperature distribution over the concrete sample during the whole heating phase. The remaining samples from group 1 to 3 were used for the tests of creep and shrinkage. The climate conditions as well as the rate of heating applied during the creep tests were different and are described in detail in Chapter 3.2.5.

3.2.2 Determination of the relative humidity and water content

Throughout the whole process of conditioning as well as during the tests of creep and shrinkage of the concrete samples, the relative humidity and temperature in the concrete pores were measured on selected samples (see sample type 3 in Fig. 3.2) equipped with 4 miniaturised probes (D/L = 4/20 mm) with a capacitive combined humidity and temperature sensor (see also Appendix A.2). The sensors were numbered according to their position to the middle axis of the concrete samples. Sensor S1 was located in the middle axis, and sensors S2, S3 and S4 were located 10, 20 and 30 mm away from the middle axis, respectively (see Fig. 3.6).

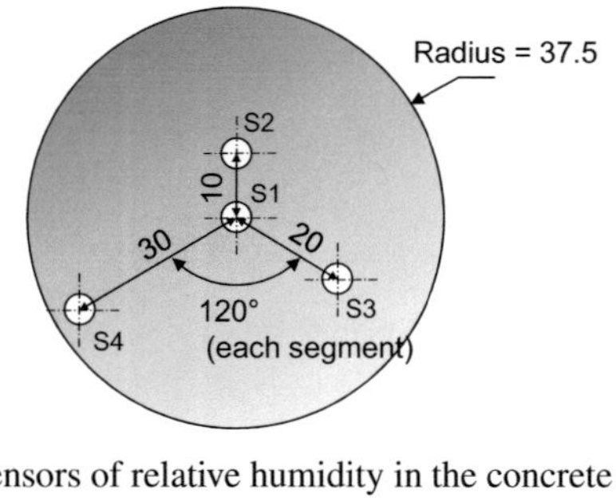

Figure 3.6: Position of the sensors of relative humidity in the concrete samples (dimensions in mm)

The development of the relative humidity in the concretes was measured in two samples per concrete mixture for the conditioning at 20 °C (see Groups 1 to 3 in Fig. 3.5) and one sample per concrete mixture for the conditioning at higher temperatures and during the tests of creep and shrinkage. Likewise, samples to measure the water content (see sample type 4 in Fig. 3.2) accompanied the different conditioning scenarios. The changes in the water content was measured periodically in at least 4 samples per concrete mixture and conditioning scenario. The dry weight of the samples was determined after drying them at 105 °C in accordance with DIN EN ISO 12570 [N10]. The results of these measurements are given in Chapter 4.1 and Appendix B.1.

3.2.3 Evaluation of the concrete microstructure

As a result of the various w/c-ratios of the mixtures, it is expected that the concretes exhibit different porosities, different specific surface areas and more important, different pore size distributions. Additionally, it needs to be verified whether the influence of temperature could also have affected the microstructure of the concretes subject to elevated temperatures by means of hydrothermal reactions between silicone dioxide from the aggregates and calcium hydroxide from the hardened cement paste (see Chapter 2.1.3). These features were investigated by means of mercury intrusion tests and gas permeability tests according to DIN 66133 [N3] and DAfStb booklet 422 [N13], respectively.

The samples required for conducting these tests were obtained from remains of samples used for tensile strength tests. From these remains, slides with a height of approximately 10 mm were cut to further extract small pieces of hardened cement paste from the edges of them and conduct the mercury intrusion tests. Before testing, the hardened cement pastes were oven-dried for 12 hours at 105 °C. The maximum intrusion pressure used was 2060 bar. The evaluation of the measurements was carried out on the basis of a contact angle of 141.3° for the oven-dried cement paste and a mercury surface tension of 0.485 N/m. Hence pore radius in the range of about 0.0037 to 170 μm could be registered. Additionally, slides with a height of 30 mm were cut to be used for the tests of gas permeability. The gas permeability was measured applying pressure in three levels; 2.0, 2.5 and 3.0 bar for the samples from the concretes MRC and MHC and 2.5, 3.0 and 3.5 bar for the samples from the concrete MLC.

Table 3.6 presents the number of samples tested according to the conditioning of relative humidity and temperature before and during heating. To compare the microstructure between the concrete mixtures, samples which were subject neither to drying nor to elevated temperatures were used. In Table 3.6, these samples correspond to the ones conditioned at 20 °C and 100 % RH. In addition to evaluating the differences of the pore size distributions between the concrete mixtures, concretes subject to high temperatures and different humidities were compared with concretes kept at ambient temperature. Based on these comparisons, indications of the influence of temperature on the concrete microstructure can be qualitatively assessed. Further validation of the results from mercury intrusion tests were achieved by conducting tests of gas permeability. By comparing the permeability coefficient of "warm" and "cold" concretes an indication of changes in the microstructure due to exposure to elevated temperature can also be identified. In case hydrothermal reactions occurred, it should be possible to recognize a densification of the concrete microstructure in both mercury intrusion and gas permeability tests.

Table 3.6: Tests conducted for the evaluation of the concrete microstructure

Concrete Mixture	Conditioning at 20 °C	Conditioning during heating			Number of samples	
	RH [%]	RH [%]	Heating		Mercury Intrusion	Gas Permeability
			T [°C]	Δt [d]		
MRC	100	-	-	-	3	3
	95	sealed	60	1~7	2	2
		95	60	60	2	2
			60	90	2	2
	85	sealed	60	1~7	2	2
		75	60	60	2	2
			60	90	2	2
			80	90	2	2
	65	65	60	30	2	2
			60	90	2	2
			80	90	2	2
MLC and MHC	100	-	-	-	3, 3 (Σ=6)	3, 3 (Σ=6)
	85	85	60	1~7	2, 2 (Σ=4)	2, 2 (Σ=4)
	Total number of samples:				33	33

The tests to evaluate the concrete microstructure were carried out on 3 samples of each concrete mixture conditioned at 20 °C and 2 samples for each concrete mixture subject to elevated temperatures. The results of these tests are given in Chapter 4.2 and Appendix B.2.

3.2.4 Determination of strength and stiffness

The tests were conducted without changing the temperature applied to the samples during the conditioning. For this purpose, a testing machine with an integrated temperature chamber was employed. For the samples tested at elevated temperatures, the temperature chamber allowed to keep the temperature during testing equal to the temperature of conditioning. However, neither at elevated temperatures nor at room temperature, it was possible to fix the relative humidity of the environment according to the relative humidity of the conditioning. Hence, all samples were sealed during testing using aluminium cored polypropylene films to prevent any moisture exchange between the samples and the environment. The material mechanical properties of tensile strength, compressive strength and modulus of elasticity of the concrete samples were investigated according to the following test methods:

- Compressive strength was measured on cylindrical samples with a diameter-height relation of 1/2 (see sample type 1 in Fig. 3.2). The force was applied at a constant rate of 0.6 N/(mm^2 s) in accordance with DIN EN 12390-3 [N8].
- For the measurements of tensile strength, samples type 2 (see Fig. 3.2) were used. Uniaxial tension was applied directly to the concrete samples at constant rate of 0.05 N/(mm^2 s) following the recommendation from DAfStb booklet 422 [N13]. In order to ensure that uniform tensile stresses predominate in the sample body, the tensile load was exerted by steel plates fixed to the testing machine. The samples were glued to the steel plates using a two-component adhesive based on methyl methacrylat (X60) with a hardening time of approximately 15 minutes at 20 °C. The samples tested at higher temperatures were glued to the plattes using a two-component epoxy adhesive (3M DP 490) with longer hardening times and high strength at elevated temperatures. This epoxy adhesive required approximately 2 hours for hardening at 60 °C and 1 hour at 80 °C.
- Modulus of elasticity was measured after loading and unloading the specimens three times with a constant rate of 0.5 N/(mm^2 s) between 0.5 N/mm^2 and 1/3 of the compressive strength following the procedure given in DIN 1048-5 [N2]. The strain measures were done using a clamp-on strain transducer based on a strain gauge measuring system. Two transducers measured the displacement of the concrete samples within a reference distance of 75 mm. After measuring the modulus of elasticity the samples were loaded until failure to determine the compressive strength. These results were also taken into account for the evaluation of compressive strength.

Table 3.7 provides an overview of the laboratory tests carried out to investigate the influence of drying and temperature on the compressive strength, tensile strength and modulus of elasticity of the concrete mixtures MRC, MLC and MHC. The tests were conducted to evaluate the different influences of temperature and water loss according to the approach described in Chapter 3.2. The reference values correspond to the results from the tests conducted at reference conditions (20 °C and 100 % RH). The influence of drying is evaluated by the tests conducted after drying the samples at reference temperature and different relative humidities (20 °C and 95, 85, 65, 0 % RH). The influence of temperature is evaluated by the tests conducted on samples that were conditioned at reference temperature and then heated up to temperatures of 60 °C or 80 °C. Some of the heated samples were sealed before heating to avoid any changes in the moisture content and then tested within a week after increasing the temperature in order to assess the influence of temperature at different water contents of the concrete. The remaining heated samples were tested after 30, 60 or 90 days of sustained elevated temperature and were allowed to exchange moisture with the surroundings at defined conditions of relative humidity (95, 75, 65 % RH).

Table 3.7: Tests conducted for the investigation of concrete strength and stiffness

Concrete Mixture	Conditioning at 20 °C RH [%]	Conditioning during heating RH [%]	Heating T [°C]	Heating Δt [d]	Number of samples Compressive Strength	Number of samples Modulus of Elasticity	Number of samples Tensile Strength
MRC	100	-	-	-	5	3	5
		sealed	60	1	4	3	4
	95	-	-	-	6	3	6
		95	60	1~7*, 60, 90	5, 5, 5 (Σ=15)	3, 3, 3 (Σ=9)	5, 3, 3 (Σ=11)
	85	-	-	-	6	3	6
		75	60	1~7*, 60, 90	5, 5, 5 (Σ=15)	3, 3, 3 (Σ=9)	5, 3, 3 (Σ=11)
			80	1~7*, 90	5, 4 (Σ=9)	3, 3 (Σ=6)	4, 3 (Σ=7)
	65	-	-	-	6	3	6
		65	60	1~7*, 30, 90	5, 5, 5 (Σ=15)	3, 3, 3 (Σ=9)	5, 3, 3 (Σ=11)
			80	1~7*, 90	5, 4 (Σ=9)	3, 3 (Σ=6)	4, 3 (Σ=7)
	0	-	-	-	4	3	4
	vacuum	-	-	-	5	3	5
		sealed	60	1~7	4	3	4
MLC and MHC	100	-	-	-	5, 5 (Σ=10)	3, 3 (Σ=6)	5, 5 (Σ=10)
	95	-	-	-	6, 6 (Σ=12)	3, 3 (Σ=6)	6, 6 (Σ=12)
	85	-	-	-	6, 6 (Σ=12)	3, 3 (Σ=6)	6, 6 (Σ=12)
		sealed	60	1~7	6, 6 (Σ=12)	3, 3 (Σ=6)	4, 4 (Σ=8)
	65	-	-	-	6, 6 (Σ=12)	3, 3 (Σ=6)	6, 6 (Σ=12)
		sealed	60	1~7	6, 6 (Σ=12)	3, 3 (Σ=6)	4, 4 (Σ=8)
	0	-	-	-	4, 4 (Σ=8)	3, 3 (Σ=6)	4, 4 (Σ=8)
	vacuum	-	-	-	5, 5 (Σ=10)	3, 3 (Σ=6)	5, 5 (Σ=10)
	Total number of samples:				191	111	167

* Samples tested within a week after reaching the desired temperature were heated in sealed conditions.

The minimum number of samples used for determining the compressive strength was 4 and for the modulus of elasticity and tensile strength was 3. The majority of the tests were conducted on the reference concrete MRC. For the concretes MLC and MHC more importance was given to the influence of drying, although the influences of temperature after drying at 20 °C and 85 % RH as well as 20 °C and 65 % RH were also measured. The results of these measurements are given in Chapter 4.3 and Appendix B.3.

3.2.5 Determination of creep and shrinkage

The tests to investigate the creep behaviour of concrete were conducted based on the recommendations given in DAfStb booklet 422 [N13] and RILEM TC 107-CSP [N19]. Prior to testing, the concrete samples were conditioned over 450 days at 20 °C and relative humidities of 95, 85 and 65 %. The samples used for measuring creep and shrinkage were of type 1 according to Fig. 3.2. Concrete creep was investigated for three climatic conditions: 20 °C / 65 % RH, 40 °C / 65 % RH and 70 °C / 65 % RH. Traditional creep test machines equipped with hydropneumatic accumulators were used for the samples tested at 20 °C, and for the samples tested at higher temperatures, creep climate chambers were developed and integrated to the creep testing machines. The features and capabilities of these climate chambers are given in Appendix A.3.

The time scheme of the creep tests performed under elevated temperatures is shown schematically in Fig. 3.7. The concrete samples were heated at a rate of 10 K/h. One hour after the chamber reached the desired temperature, the samples were loaded mechanically for a period of 42 days and then unloaded while the measurements continued 21 days longer to appraise the reversibility of creep. The load was set in every case to 1/3 of the compressive strength at the age and temperature of loading.

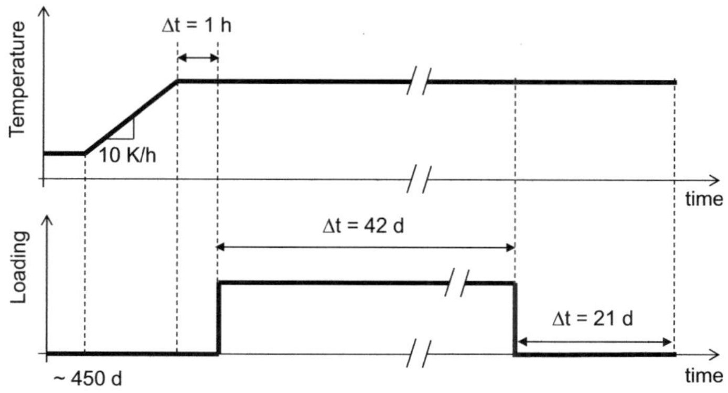

Figure 3.7: Sequence of heating and loading in the creep tests

For the selection of the heating process the following considerations were evaluated:

- As shown by Budelmann [35] (see also Rostásy and Budelmann [128]), the measured values of creep are highly dependent on the heating regime. Imposing the load after heating leads to considerable lower creep deformations than those that are obtained in samples loaded before increasing the temperature. It is well known that the creep deformations are highly dependent on the moisture gradient and moisture movement within the concrete samples [98]. During heating a rash movement of water molecules from the gel pores to the capillary pores takes place leading to an increase in the relative humidity of the pore cavities (see Chapter 2.1.2, Bažant [6]). The influence of this moisture movement on the concrete creep is the source of the differences of creep deformation between the heating regimes described by Budelmann (transient creep) and can only be measured if the load is applied before the temperature is increased. Nevertheless, the interpretation of the measurements for the process of loading before heating presents some difficulties. During heating, the loaded samples suffer creep, thermal expansion, and shrinkage deformations at the same time and the unloaded samples show thermal expansion and shrinkage which cannot be appraised separately. Additionally, the measure device is also affected by thermal expansion which leaves the measurements without any reference from which an absolute reliable value of deformation can be derived.

- For the case that the load is applied after increasing the temperature, any influence of the temperature on the measure device during heating can be neglected because the measure device reaches the desired testing temperature before the measurements of creep and shrinkage that are relevant for the investigation begin. During the process of heating, neither the influence of moisture movement between gel and capillary porosity nor the influence of any exchange of moisture with the environment on the creep deformations can be recorded because throughout this process the samples remain unloaded. However, the influence of moisture movement can be estimated afterwards by comparing the measurements of relative humidity within the concrete samples with the corresponding creep and shrinkage deformations and the influence of moisture loss during heating can be minimized by reducing the duration of the heating period i.e. increasing the heating rate.

- Due to the good thermal conductivity of the concrete, setting the heating rate of the surrounding air to 10 K/h does not induce any considerable temperature gradients within the concrete samples during heating. However, the heat exchange between concrete samples and air within the temperature chambers had to be improved in order for the concrete samples to reach the desired temperature within the provided time period. Natural convection is not enough to reach this goal, therefore, air blowers were installed in the climate chambers.

The use of air blowers produced some consequences for the realisation of the test. First, it forces the air to circulate, which ensures a constant temperature and a constant relative humidity distribution of the air in the chamber. Second, it increases the heat transfer capabilities of the air surrounding the samples. The air flew inside the chamber with a velocity of approximately 0.7 m/s allowing the samples to reach, at the moment of loading, a mean temperature between 2 to 3 °C below the desired temperature and the desired temperature about 24 hours later. Third, it may have accelerated the process of drying of the samples. The first two consequences constitute the intention of using air blowers, while the third consequence though unintended, do not represent any harm to the validity of the measurements because the drying process of the samples was assessed by relative humidity sensors. Some preliminary tests conducted with the climate chambers which support the necessity of using air blowers are presented in Appendix A.4.

Two creep samples and two unloaded samples were measured in every creep test carried out for the reference mixture MRC. After the creep tests on the concrete MRC were completed, it was decided to simplify the creep tests of the concretes MLC and MHC by using only one creep sample and one shrinkage sample for every test. This decision was supported on the reproducibility of the results shown by the measurements conducted on the concrete MRC. The number of concrete samples employed are listed in Table 3.8 according to the conditions of temperature and relative humidity of the creep and shrinkage tests.

Table 3.8: Tests conducted for the investigation of concrete creep

Concrete Mixture	Conditioning at 20 °C	Conditioning during loading		Number of samples	
	RH [%]	RH [%]	Temperature [°C]	Creep	Shrinkage
MRC	65	65	20	2	2
			40	2	2
			70	2	2
	85	65	20	2	2
			40	2	2
			70	2	2
	95	65	20	2	2
			70	2	2
MLC and MHC	65	65	20	1, 1 (Σ=2)	1, 1 (Σ=2)
			70	1, 1 (Σ=2)	1, 1 (Σ=2)
	85	65	20	1, 1 (Σ=2)	1, 1 (Σ=2)
			70	1, 1 (Σ=2)	1, 1 (Σ=2)
	Total number of samples:			24	24

For each concrete sample, the creep deformations were measured using three inductive displacement transducers mounted on rings made of stainless steel. The rings were fixed to the samples in three points by means of fastening screws. The distance of reference between the rings for the calculation of the strains was 100 mm. Due to lack of space within the creep climate chambers, the deformation of the shrinkage samples was measured using only two inductive displacement transducers for each sample. The results of the creep and shrinkage measurements are given in Chapter 4.4 and Appendix B.4.

3.3 Summary

Based on the literature review done in Chapter 2 and in particular on the experimental investigations conducted by Budelmann [35], it can be stated that the changes in the mechanical properties of concrete subject to elevated temperatures are decisively influenced by the moisture content of the concrete. The conception of the experiments presented in this chapter is based on the assumption that the influence of temperature on the mechanical properties of concrete can be explained by superposition of the effects that temperature and moisture content cause on the concrete mechanical properties.

The experimental program is focused on investigating the impact of different climate conditions on strength, stiffness and time dependent deformations of concrete. The experiments evaluate the influence of moisture content solely, the influence of increasing temperature at different moisture contents, and the influence in time of a sustained elevated temperature on the compressive strength, tensile strength and modulus of elasticity of concrete. In addition, the influence of different temperatures and moisture contents on the time dependent deformations of concrete is as well evaluated.

Analysing the results of the tests described in the experimental program serves to improve the understanding of the different factors that play a role in the influences of temperature on the concrete mechanical properties. With the acquired knowledge and an exhaustive evaluation of the tests results, it shall be possible to produce models that can accurately predict the mechanical behaviour of concrete at different temperatures and humidity conditions.

Chapter 4

Experimental results and discussion

The analysis of the results from the experimental investigation allows to properly understand the different effects that are involved in the influence of variable conditions of temperature and humidity on the mechanical properties of concrete. This chapter presents the most relevant findings of the experimental investigation including a rigorous discussion and interpretation of the results. Because of the extensiveness of the experimental program, not all the results of the experimental investigation are presented in this chapter. The results included are used as guide to explain the process of conception of the new material model proposed in the following chapters. The basis of several assumptions required for the formulation of the models are justified according to the interpretation of the results presented here. Additional results which are not included in this chapter but support these assumptions as well, are mentioned in this chapter and included in Appendix B.

4.1 Development of water content and relative humidity in concrete by drying

As mentioned in Chapter 3.2.2, the relative humidity (RH) and temperature in the concrete pores were measured on selected samples using miniaturised sensors. These humidity sensors guarantee a high accuracy of about ± 2 % in the ranges between 5 % and 95 % RH. For relative humidities close to 100 % the sensors lose in accuracy to approximately ± 5 % (see Appendix A.2). For this reason, in the samples stored at 95 % RH it was not possible to accurately measure the development of the relative humidity. The sensors showed a constant relative humidity of 100 % in all concrete mixtures for measurements conducted both at reference temperature as well as at elevated temperatures.

The concrete samples used to measure water content were small in relation with the maximal particle size of the aggregates (see sample type 4 in Chapter 3.1.3). This fact affects the homogeneity of the samples and therefore the possibility of comparing them with each other directly. A comparison between the samples is therefore only possible if the measured values of water content are normalised to a reference value. The development in time of the water loss measured in the samples is thus, in the following sections, presented as a percentage of the original amount of water stored in the concrete samples before the beginning of drying and will be referred to as normalised values of water content. The following sections present some results of the measurements of water content and relative humidity from samples drying at room temperature (20 °C) as well as at elevated temperatures. Further measurement results are presented in Appendix B.1.

4.1.1 Concretes drying at room temperature

Fig. 4.1 presents the normalised values of water content from the concrete mixtures MLC (w/c = 0.4), MRC (w/c = 0.5) and MHC (w/c = 0.6) drying at 95 %, 85 % and 65 % RH during a period of about 700 days. The symbols of the legend represent measurements and the lines connecting them are used to facilitate the reading of the diagram. As described in Chapter 3.2.1, from the group of samples subject to 20 °C and 65 % RH some samples were moved and stored at an environment of 20 °C and 0 % RH during 150 days and finally subject to vacuum for 100 days longer. The normalised values of water content of these samples are also presented in Fig. 4.1. The results are predictable and consistent. By a given concrete mixture, the lower the relative humidity of the environment, the higher the water loss from the concrete samples. Furthermore, by a given relative humidity of the environment, the lower the w/c-ratio of the concrete samples the slower the development of the water loss in time.

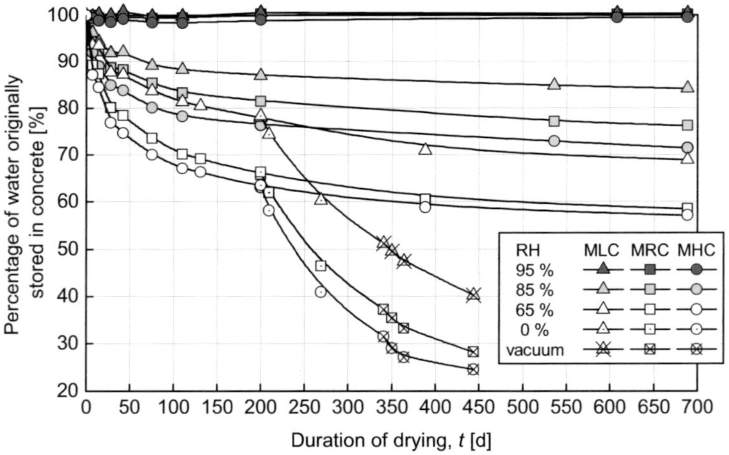

Figure 4.1: Time development of water stored in concrete in relation to the water stored before the beginning of drying for the concretes MLC, MRC and MHC drying at 20 °C and different ambient relative humidities

As mentioned before, the sensors of relative humidity measured a constant value of 100 % RH by the samples drying at 95 % RH and the water content of these samples did not show any significant reduction (see Fig. 4.1). Based on these facts and on the evidence that the sensor are inaccurate at high humidities, it can be stated that the relative humidity in the pore cavities of the samples stored at 95 % RH and 20 °C was maintained very close to 100 %. Therefore, in further it is assumed for the interpretation of the results, that these concrete samples reached a relative humidity of 99 % at the end of the drying process at 95 % RH.

4.1.1.1 Influence of the environmental relative humidity

Fig. 4.2 shows the measurements of the moisture distribution in the concrete samples made with the mixture MRC and stored at 20 °C with 85 % and 65 % RH. The measurements of relative humidity were conducted at intervals ranging from 10 minutes to 4 hours. However, for sake of simplicity in the following figures only some selected measurements are shown represented by the symbols of the legend. In addition, lines connecting the symbols are used to facilitate the reading of the diagram. The layers of the legend are named according to the position of the sensors as described in Chapter 3.2.2: sensor S1 was placed in the middle axis of the sample, and sensors S2 to S4 are located at distances of 10, 20 and 30 mm from the middle axis of the sample, respectively. The mean values of relative humidity in the concrete samples were calculated in every case by calibrating a function that could fit the measured results and approximate the humidity profile of the samples in the best way. The process followed to obtain the mean values of relative humidity in the concrete samples is described in Appendix B.1.

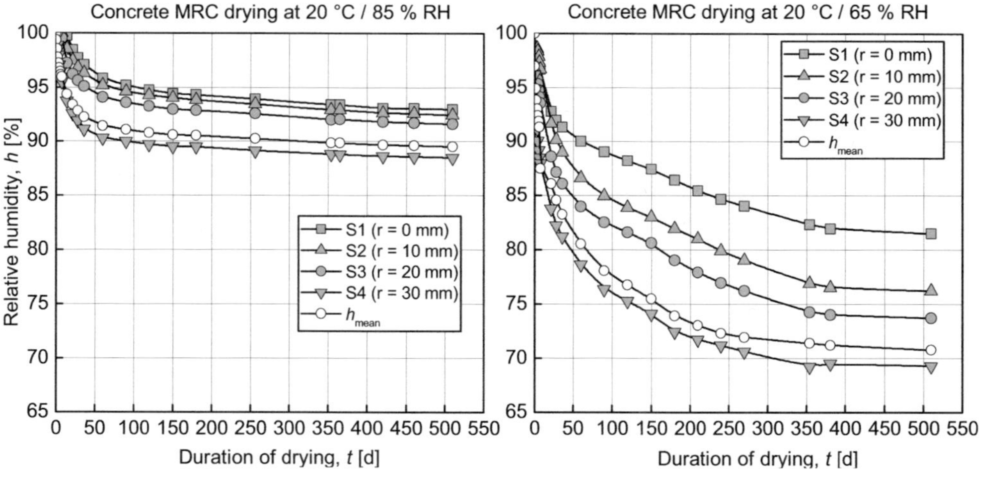

Figure 4.2: Measurements of relative humidity in the concrete MRC during drying at 20 °C / 85 % RH (left) and 20 °C / 65 % RH (right)

In Fig. 4.2, it can be clearly seen that even after a drying period of about 500 days, an equilibrium between the relative humidity in the concrete pores and the relative humidity of the surrounding air could not be reached. This is also confirmed by the measurements of moisture content presented in Fig. 4.1 in which the curves of the concretes drying at 85 % and 65 % RH still show a significant gradient after 500 days.

According to the model of moisture transport from Bažant and Najjar [10, 11] and also from Kießl and Gertis [81], as well as experimental investigations conducted by van der Kooi [87], the water diffusion coefficient is dependent on the water content (see also Chapter 2.2.3). The higher the water content, the easier the water molecules can diffuse through the concrete body, which implies that concrete samples drying at higher environmental relative humidity would reach the equilibrium between relative humidity of the pores and the surrounding faster than those drying at lower relative humidities. This approach is according to the measurements presented in Fig. 4.2, not quite correct. After 500 days of drying the concrete MRC stored at 85 % RH had a mean relative humidity of 89.5 %, which corresponds to 70 % of the path between 100 % and 85 % RH. Meanwhile, the concrete MRC stored at 65 % RH had a mean relative humidity of 70.8 % after 500 days, which corresponds to 83 % of the path between 100 % and 65 % RH. This behaviour depicts that at high environmental relative humidity the concrete does not reach its equilibrium with the environment faster than when subject to low relative humidities. This tendency was also seen on the measures conducted on the concretes MLC and MHC which are given in Appendix B.1. On this background, the dependency of the water diffusion coefficient of concrete on the moisture content shall be re-evaluated. It will be shown in Chapter 6.1.1 that, in order to achieve a better agreement between simulations and measurements, the dependency of the water diffusion coefficient on moisture should be ruled by the environmental relative humidity, which defines the final value to reach, and the gradient of the relative humidity.

4.1.1.2 Influence of the concrete microstructure

The influence of the microstructure on the drying process of concrete can be recognised in Fig. 4.3. This figure compares the behaviours of the concretes MLC (w/c = 0.4) and MHC (w/c = 0.6) drying at 20 °C and 65 % RH. It is expected that a concrete with less capillary porosity exerts higher resistance to the movement of the water molecules within it. The measurements agree with this expectation very well. According to Fig 4.3, reaching the environmental relative humidity will take much longer for the concrete MLC than the concrete MHC. After 500 days of drying by 65 % RH, the concrete MLC had a mean relative humidity of 78.9 %, while the concrete MHC lost much more moisture to the surroundings in the same period of time reaching a mean relative humidity of 70.7 %.

In comparison with the concretes MRC and MHC, the concrete MLC dries much slower. This was seen in the measurements of moisture content as well as relative humidity (see Figs. 4.1 to 4.3). However, even when the concrete MRC (w/c = 0.50) has less capillary porosity than the concrete MHC (w/c = 0.60), the measurements of relative humidity show a similar behaviour between these two concretes by drying at an ambient relative humidity of 65 % (compare Figs. 4.2 and 4.3) as well as at 85 % (see Appendix B.1).

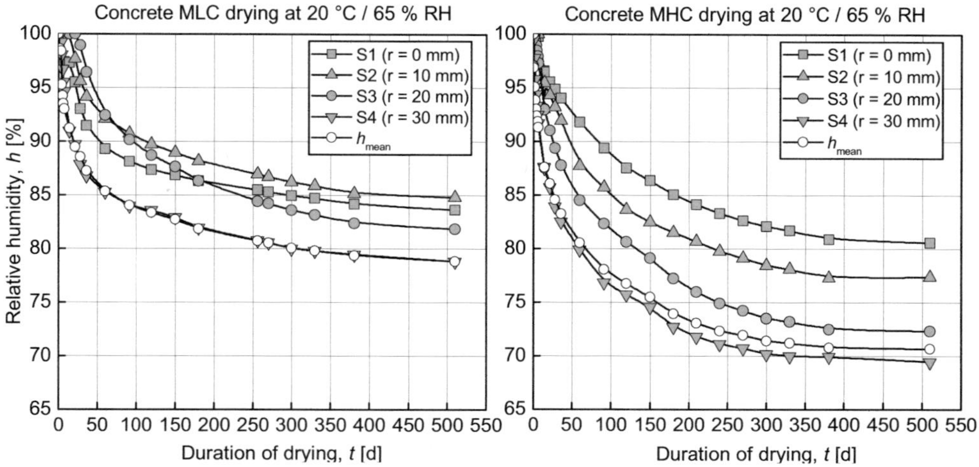

Figure 4.3: Measurements of relative humidity in the concretes MLC (left) and MHC (right) during drying at 20 °C / 65 % RH

Nevertheless, if the measurements of water loss are considered, a similar behaviour between concretes MRC and MHC cannot be seen. As shown in Fig. 4.1, by drying at environmental relative humidities of 85 % and 65 % the concrete MHC shows a faster loss of moisture than the concrete MRC. This behaviour is however consistent with the water store capacity of the concretes. The sorption isotherms in concretes with higher capillary porosity show higher gradients in the ranges close to 100 % RH due to the greater amount of capillary pores in their microstructures (see e.g. Espinosa [49], Kielsgaard [83]). The ranges close to 100 % RH contain therefore a more important portion of the total stored water for concretes with higher w/c-ratios. It can then be stated that after reaching a given relative humidity, the percentage of water originally stored for concretes with higher w/c-ratio is lower than for concretes with lower w/c-ratios.

4.1.2 Concretes drying at elevated temperatures

Fig. 4.4 presents the normalised values of water content from the concrete mixture MRC drying at 20, 60 and 80 °C and different relative humidities. The samples drying at higher temperatures were previously subject to drying at 20 °C following the sequence of conditioning presented in Chapter 3.2.1. At elevated temperatures the samples lost a large additional amount of water to the environment. For the samples previously stored at 20 °C and 95 % as well as 65 % RH the increase of temperature took place without changing the relative humidity of the environment. But even in these cases the samples lost an important portion of their water content. This behaviour can be understood by comparing sorption isotherms of the material at different temperatures. As described in Chapter 2.1.2.2, at higher temperatures the capacity of storing water in concrete reduces, which implies that at a given value of relative humidity, the water content of the concrete with the lower temperature is higher. Moreover, increasing temperature leads to an increase of the relative humidity in the concrete pores, which promotes the expulsion of water from the concrete microstructure. The higher the temperature increase the lower the amout of water allowed to remain in the concrete microstructure. The measurements of water content at elevated temperatures conducted on the concretes MLC and MHC are presented in Appendix B.1.

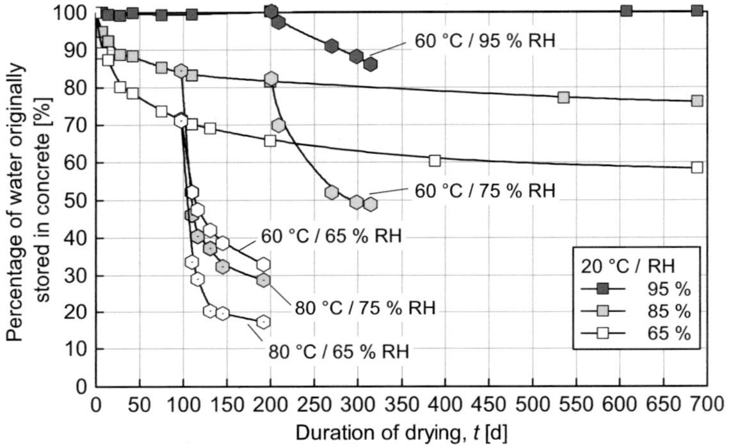

Figure 4.4: Time development of water stored in concrete MRC in relation to the water stored before the beginning of drying for temperatures of 20, 60 and 80 °C and different ambient relative humidities

Especially consideration shall be given to the samples drying at 60 °C and 95 % RH. As mentioned at the beginning of this chapter, the sensors could not identify any changes in the relative humidity of the samples drying at 95 % RH. However, as it can be seen in Fig. 4.4, after 120 days of drying at 60 °C and 95 % RH the samples lost about 15 % of the water content. This behaviour can as well be explained by the properties of water storage of the concretes and its temperature dependency. Measurements conducted by Hundt and Kantelberg [77] showed that not only the sorption isotherm of a "warm" concrete lies under the one of a "cold" concrete but also the storage capacity by relative humidities around 100 % reduces considerably when the temperature is increased.

4.1.2.1 Influence of temperature and environmental relative humidity

The following diagrams illustrate the influences of temperature and relative humidity of the environment on the drying process of the concrete samples. The beginning of the time axis corresponds to the beginning of the heating process. The samples were previously stored at 20 °C and relative humidities of 65, 85 and 95 % for approximately 100 or 200 days (see conditioning sequence in Chapter 3.2.1) which implies that the relative humidity of the samples before heating was lower than 100 %. Owing to the fact that an increase of the temperature leads to an increase of the relative humidity in the concrete pores (see Chapter 2.1.2.2), to avoid misinterpretation, the following diagrams present the measured values of relative humidity after the maximum values were reached. These maximum values were measured within 24 hours after achieving the desired temperature.

Fig. 4.5 shows the time development of the relative humidity in the pore cavities of the concrete MRC drying at 80 and 60 °C and a relative humidity of 65 %. Both samples started with nearly the same water content before being heated, however, after increasing the temperature, the water stored within the samples behaved differently. The concrete samples heated at 80 °C lost water much more rapidly. Starting by a mean value of relative humidity of 85 %, it came down to about 72 % after only 10 days of drying, while for the samples heated at 60 °C the mean relative humidity went from 87 % to 80 % in the same period of time.

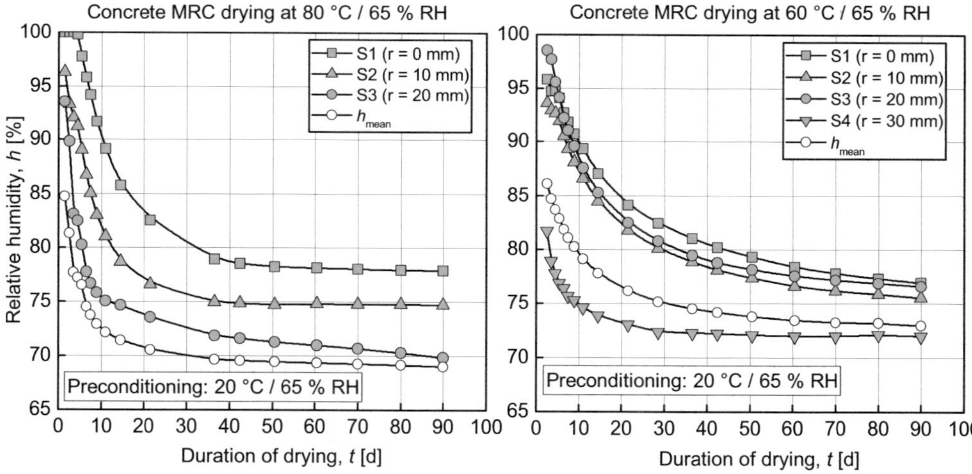

Figure 4.5: Measurements of relative humidity in the concrete MRC during drying at 80 °C / 65 % RH (left) and 60 °C / 65 % RH (right)

The relevance of these differences is more evident if the water contents are considered. According to Fig. 4.4 after 10 days of heating, the samples heated at 80 °C lost 38 % of the water content while the samples heated at 60 °C lost only around 19 %. These results provide evidences strong enough to confirm that the exposure to higher temperatures facilitates the drying of concrete members.

In Fig. 4.6 the influence of the environmental relative humidity can be explored. Samples from the concrete MHC previously stored at 20 °C / 85 % RH (left) and 20 °C / 65 % RH (right) were heated up to 60 °C at a relative humidity of 75 and 65 % respectively. Being the increase of temperature in both cases the same, the differences in the behaviour of the drying process between the two concrete samples lie on the initial water content of the samples and its interaction with the relative humidity of the environment.

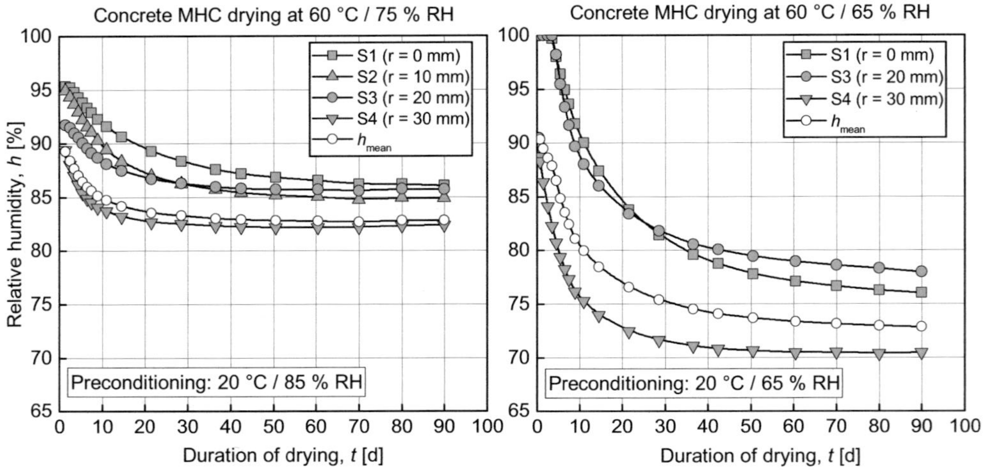

Figure 4.6: Measurements of relative humidity in the concrete MHC during drying at 60 °C / 75 % RH (left) and 60 °C / 65 % RH (right)

The concrete sample drying at 75 % RH lost less water than the sample drying at 65 % RH. The mean relative humidity of the sample drying at 75 % RH decreased about 5 % within the initial 10 days while the mean relative humidity of the sample drying at 65 % RH dropped 10 % in the same period of time. The results of the measurements conducted at elevated temperatures as well as the measurements conducted at 20 °C (see Fig. 4.2) suggest that increasing the gradient of relative humidity within the concrete samples improves the drying process of concrete.

4.1.2.2 Influence of the concrete microstructure

Fig. 4.7 illustrates the drying process of the concretes MLC (w/c = 0.40) and MHC (w/c = 0.60) previously stored at 20 °C and 85 % RH and then heated up to 60 °C at a relative humidity of 75 %. Analogous to the results of the measurements conducted at 20 °C (see Fig. 4.3), the concrete with higher capillary porosity dries faster. However the influence of the microstructure seems to be less accentuated when the concretes are compared at higher temperatures.

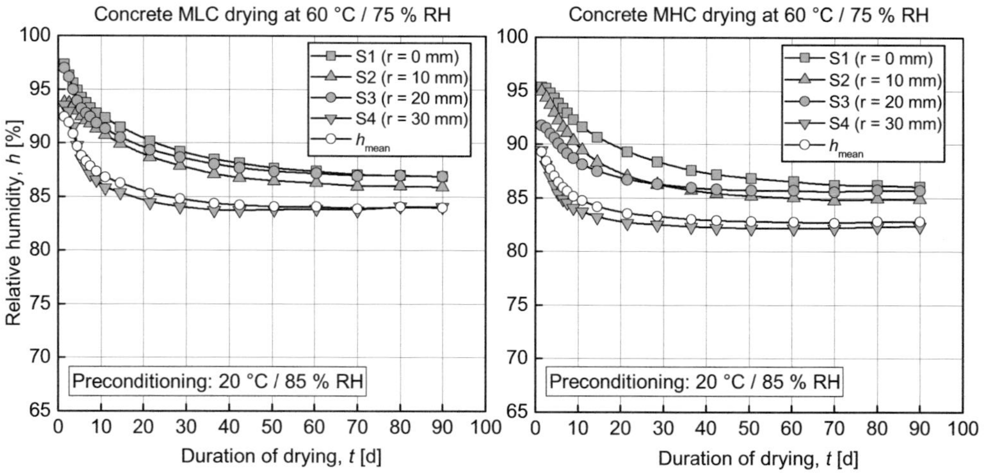

Figure 4.7: Measurements of relative humidity in the concretes MLC (left) and MHC (right) during drying at 60 °C / 75 % RH

Water within the concrete microstructure can move more easily when the concrete is subject to higher temperatures. This fact tends to wane the differences in the water transport properties between concretes with different microstructures. While the measurements conducted at 20 °C (see Fig. 4.3) show a significant difference between the drying processes of the concretes MHC and MLC, at 60 °C the difference of the values of relative humidity reached by the concretes are less meaningful (see Fig. 4.7 and Appendix B.1).

4.1.3 Temperature changes

In addition to the study of the drying process of concrete, the influence of temperature on the relative humidity of the concrete pores at constant water content was investigated. The hygrothermic coefficient of concrete (see Chapter 2.1.2.2) indicates that at constant water content, the relative humidity in concrete pores increases if the temperature increases. This effect can be appraised by measuring the relative humidity in sealed concrete samples at different temperatures. For this purpose, relative humidity measurements were carried out on samples previously dried at 20 °C and 65 % RH and then sealed and heated up in three steps to 40, 60 and 80 °C. The temperature was increased gradually until an increment of 20 °C was achieved. Then the temperature was held constant until the measured relative humidities reached constant values. This process was repeated until all three temperature steps were accomplished. The results of these measurements are presented in Fig. 4.8. In the abscissa axis the relative humidities measured in the concrete before increasing the temperature are given and the ordinate axis indicates the relative humidities reached after increasing the temperature by 20 °C.

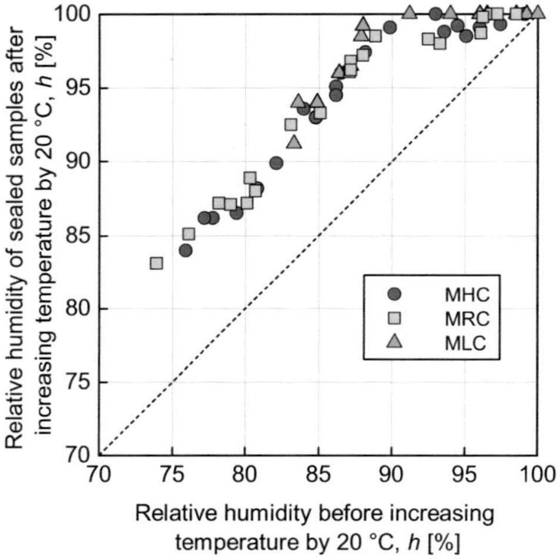

Figure 4.8: Measurements of relative humidity in the concretes MLC, MRC and MHC after incrementing the temperature by 20 °C

The dashed line in Fig. 4.8 divides the diagram in two zones. Measurements located on the line indicate no change in the relative humidity of the concrete pores due to temperature changes. Those measurements located above the line represent increments in the relative humidity due to temperature increments, and any measurement located below the dashed line would mean that the relative humidity of the concrete pores decreased after increasing temperature. As it can be seen in Fig. 4.8, incrementing the temperature by 20 °C led always to an increment in the relative humidity of the concrete pores. In the range between 75 % and 90 % RH, raising the temperature leads to an increment of the relative humidity in the concrete pores of around 8 %. At relative humidities higher than around 90 % the relative humidity increases up to the limit value of 100 % RH. With the help of these measurements an accurate mathematical formulation to calculate the influence of a temperature change on the relative humidity of the concrete pores can be proposed. In this regard, an equation to characterize the hygrothermic coefficient of concrete is derived in Chapter 6.1.3.

4.2 Measurements related to concrete microstructure

The measurements of mercury intrusion porosimetry and gas permeability were conducted with the purpose of identifying possible changes in the concrete microstructure due to the combined effects of moisture and temperature. The main goal of these measurements was to find enough evidence to suggest that under certain circumstances of moisture and temperature hydrothermal reactions between silicone dioxide from the aggregates and calcium hydroxide from the hardened cement paste took place (see Chapter 2.1.3). The hardened cement pastes used for the mercury intrusion porosimetry tests were extracted from the edge of the concrete samples (see Chapter 3.2.3), which implies that the relative humidity in their pore spaces most likely corresponds to the relative humidity of the environment at which the samples were subject. Given that the distributions of the relative humidity within the concrete samples were not constant, the results of the mercury intrusion porosimetry tests are not representative for the whole geometry of the samples. They take into account the effect of temperature and moisture of the samples at their edges. This helps however to identified at which relative humidity conditions the hydrothermal reactions took place. A more generalized indication of changes in the concrete microstructure can be appraised in the gas permeability tests. Given that the slides cut to conduct these tests include the entire geometry of the concrete samples (see Chapter 3.2.3), the results of the gas permeability tests are able to indicate changes in the concrete microstructure influencing the complete geometry of the samples.

4.2.1 Mercury intrusion porosimetry

The following diagrams show results of mercury intrusion porosimetry tests conducted on the concrete samples. In order to facilitate the interpretation of the results, ranges indicating the type of pores according to the pore size distribution proposed by Setzer [140] (see also Chapter 2.1.1.1) are labelled on the top of the figures. Fig. 4.9 presents the differential intrusion curves of the concretes MLC, MRC and MHC conditioned at 20 °C and 100 % RH.

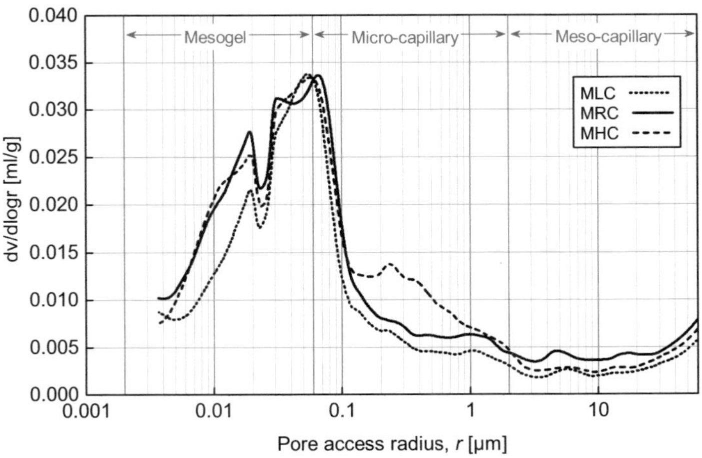

Figure 4.9: Pore size distributions of the concretes at reference conditions

Since the samples from 4.9 were subject neither to drying nor to elevated temperatures, they are considered as reference for further comparisons. Having concrete MLC the lowest w/c-ratio, it is expected to show the lowest porosity, specially in the range of capillary pores. This assumption can be corroborated by the measurements presented in Fig. 4.9. The curve of concrete MLC represented by the dotted line lies under the curves of the concretes MRC and MHC in the ranges of micro and meso-capillary pores. The differences in the microstructure of the concretes MRC and MHC are however less easy to identify. While concrete MRC shows a higher content of meso-capillary pores, the content of micro-capillary pores of the concrete MHC is clearly higher and in the range of mesogel pores, the curves of both concretes are very similar.

Fig. 4.10 provides information about the influence of temperature and moisture content on the microstructure of the concrete MRC after being dried at 20 °C and 95 % RH and then heated up to 60 °C. The dotted line corresponds to the measurements conducted on samples sealed before the temperature was increased. As mentioned in Chapter 3.2.4, these samples were conceived to appraise the effect of temperature alone avoiding any influence of drying after increasing temperature. However, as it can be seen in Fig. 4.10 in comparison with the reference curve (continuous line), the dotted line lies above in the range of mesogel pores and below in the range of micro and meso-capillary pores which is an indication of a densification of the concrete microstructure possibly due to hydrothermal reactions.

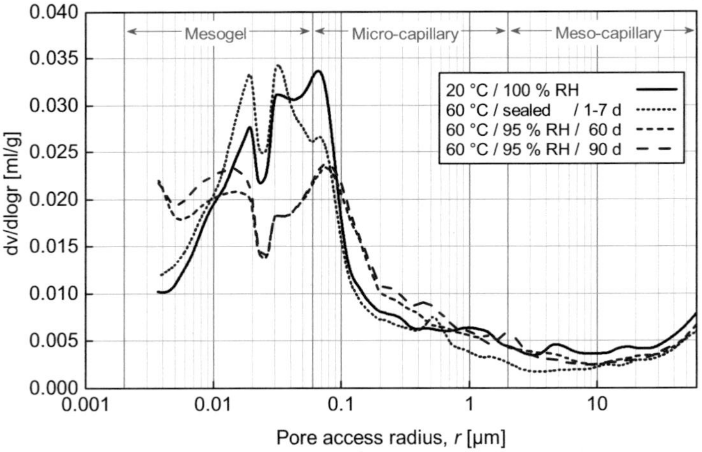

Figure 4.10: Pore size distribution of concrete MRC dried at 20 °C / 95 % RH and then subject to 60 °C / 95 % RH

Densification of the concrete microstructure was also seen in the samples that were let free to dry at 60 °C and 95 % RH represented by dashed lines in Fig. 4.10. After 60 and 90 days of drying, the measurements show an increase in the smallest pore sizes while in range of the meso-capillary pores the dashed curves lie below the curve of reference. Therefore, it can be stated that the samples heated while having a high amount of water in their microstructure were not only influenced by the effect of temperature alone but also by the combined effect of temperature and moisture in form of hydrothermal reactions.

Fig. 4.11 presents the results of the tests conducted on samples that were previously dried at 20 °C / 85 % RH and then subject to 60 °C / 75 % RH. Analogous to the results previously discussed, the sealed samples (dotted line) show a possible densification of the microstructure as evidenced by the increment of the mesogel pores and reduction in the range of meso-capillary pores when comparing with the reference curve (continuous line). This is not the case of the samples drying at 60 °C / 75 % RH (dashed lines). Possibly due to the lack of water, hydrothermal reactions did not develop. The dashed curves are displaced to the right which implies an increase of the porosity mostly in the range of micro-capillary pores and the small mesogel pores vanished.

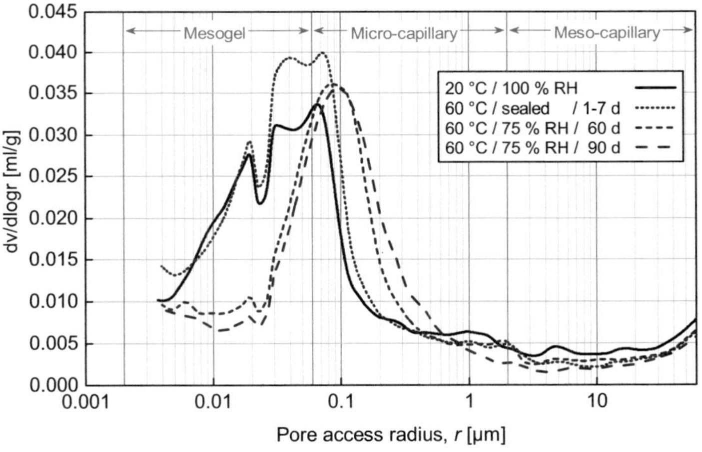

Figure 4.11: Pore size distribution of concrete MRC dried at 20 °C / 85 % RH and then subject to 60 °C / 75 % RH

The microstructural changes suffered by samples previously dried at 20 °C / 65 % RH can be evaluated in Fig. 4.12. Increments of the micro-capillary pores and vanishing of the small mesogel pores were measured in the samples after being subject to drying at 60 °C / 65 % RH during 30 and 90 days. Any development of hydrothermal reactions at such low relative humidities can be discarded. In comparison with the samples previously dried at 20 °C / 85 % RH (see Fig. 4.11), the displacement to the right of the curves from the samples previously dried at 20 °C / 65 % RH is less significant which may imply that higher content of water in the concrete leads to higher deterioration of its microstructure. This interpretation must however be confirmed by the measurements of mechanical properties.

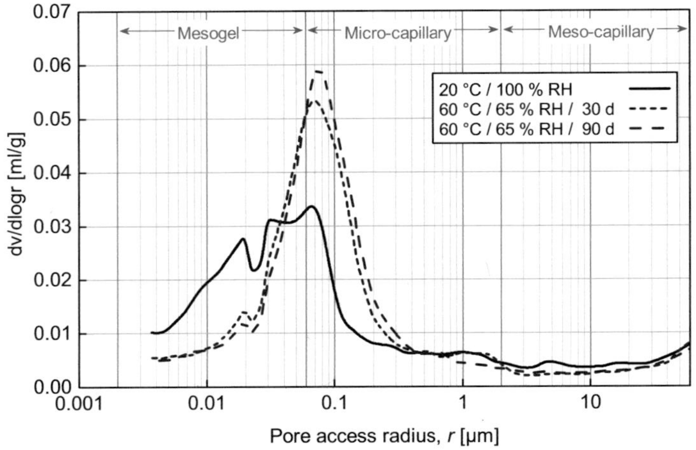

Figure 4.12: Pore size distribution of concrete MRC dried at 20 °C / 65 % RH and then subject to 60 °C / 65 % RH

The results of the mercury intrusion porosimetry tests show that samples containing a high amount of moisture (above 75 % RH) that were sealed and then heated during 1 to 7 days as well as samples that dried at very high relative humidity (95 %) during 60 and 90 days experienced a densification of their microstructure due to hydrothermal reactions. These results only identify whether such reactions took place or not and do not allow to draw any conclusion about the time-development of the hydrothermal reactions. Complementary measurements conducted on the concretes MRC at 80 °C as well as on the concretes MLC and MHC are presented in the Appendix B.2.

4.2.2 Gas permeability

Tests of gas permeability were conducted in order to complement the results of the measurements of mercury intrusion porosimetry and corroborate the evidences that those measurements provided in regard of densification of the concrete microstructure through hydrothermal reactions. As described in Chapter 3.2.3, the concrete slides used for testing gas permeability and mercury intrusion porosimetry were obtained from the same samples. Therefore, the conclusions drawn from the mercury intrusion porosimetry tests shall be validated by the gas permeability tests.

It is well known that the w/c-ratio establishes the size and amount of capillary pores within the concrete microstructure which are essential to configure the permeability of the concretes. The permeability coefficients measured from samples of the concretes stored at reference conditions (20 °C / 100 % RH) are presented in Table 4.1. These results follow the same tendency seen in the measurements of relative humidity and mercury intrusion porosimetry. The lower the w/c-ratio the denser the microstructure of the concretes and therefore the lower the permeability.

Table 4.1: Permeability coefficients of the concretes at reference conditions

Concrete Mixture	MLC	MRC	MHC
Permeability coefficient, K_{ref} [10^{-17} m²]	0.90	1.88	7.10

A comparison between the values of permeability after subjecting the concretes to different elevated temperatures and relative humidities with the values of permeability measured from samples at reference conditions can as well suggest changes in the concrete microstructure. Lower permeability coefficients are related to densification and higher permeability coefficients to damage of the concrete microstructure. Fig. 4.13 shows the permeabilities measured on samples of the concrete MRC after being subject to 60 °C and several ambient humidities in relation to the permeability obtained from samples tested at reference conditions. Three different hatches are used to identify the conditions of the samples before the heating took place.

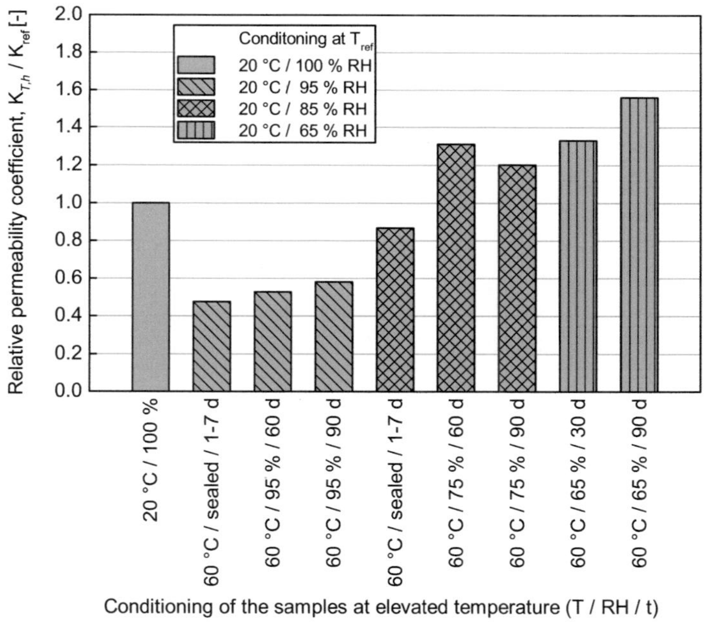

Figure 4.13: Permeability coefficients of concrete MRC subject to different temperature and humidity conditions in relation to the permeability coefficient measured at reference conditions

The permeability coefficient of the samples previously stored at 20 °C / 95 % RH decreased between 50 and 40 % after being subject to 60 °C and 95 % RH which means that the condition of high temperature and humidity endorsed the densification of the concrete microstructure. These results agree with the results from mercury intrusion porosimetry tests presented in Fig. 4.10. Likewise, the results presented in Fig. 4.11 agree with the behaviour of the permeability coefficients shown by the samples previously stored at 20 °C / 85 % RH and then subject to 60 °C and 75 % RH. Densification of the microstructure was only seen by the samples heated in sealed conditions while the samples heated in an environment with 75 % RH showed an increment of the permeability coefficients. Finally, the samples previously store at 20 °C / 65 % RH and then heated up to 60 °C at 65 % RH suffered damage in the microstructure distinguished by an increment of around 40 % in the permeability coefficients. This behaviour is again in accordance with the measurements of mercury intrusion porosimetry (see Fig. 4.12). Further measurements of permeability included in Appendix B.2 from concrete MRC heated at 80 °C and concretes MLC and MHC heated at 60 °C also show concordance with the results of the mercury intrusion porosimetry tests.

The results of the tests conducted to evaluate the effects of temperature in the concrete microstructure proved that concretes that were heated while containing a relative high amount of moisture (above 75 % RH) suffered structural changes that led to a densification of their microstructures. These changes can probably be originated by the development of hydrothermal reactions between silicone dioxide of the aggregates and calcium hydroxide from the cement paste. Hence, assuming that the influence of temperature on the mechanical properties of concrete can be explained by superposition of the effects caused by temperature and moisture content do not apply to concretes with high moisture contents. The principle of superposition applies for linear systems where the variables are linearly independent. The samples heated while having a high amount of water in their microstructure were not only influenced by the effect of temperature alone but also by the combined effect of temperature and moisture in form of hydrothermal reactions which can be interpreted as a non linearity added to the physical system.

4.3 Measurements of concrete strength and stiffness

The effects of temperature and moisture content on the strength and stiffness of concrete were investigated. The resulting mean values and standard deviations are given in Table 4.2. Highlighted in bold typeface are the mean values obtained from the samples tested at reference conditions (20 °C / 100 % RH) which correspond to the reference values. The quotient between the mean values from measurements conducted on samples subject to different ambient conditions and these reference values are used to quantify the influences of moisture and temperature on the mechanical properties of the concretes.

Table 4.2: Mean values and standard deviations of compressive strength, tensile strength, and modulus of elasticity by the investigated conditions of temperature and moisture

Concrete Mixture	Conditioning at 20 °C RH [%]	Conditioning during heating RH [%]	Heating T [°C]	Heating Δt [d]	Mean values f_{cm} [N/mm²]	f_{ctm} [N/mm²]	E_{cm} [N/mm²]
MRC	**100**	-	-	-	**51.5 ± 4.1**	**3.64 ± 0.08**	**40939 ± 2504**
		sealed	60	1	45.5 ± 3.3	2.60 ± 0.41	26539 ± 210
	95	-	-	-	56.6 ± 1.9	3.97 ± 0.63	32230 ± 1734
		sealed	60	1~7	54.2 ± 3.5	3.72 ± 0.51	29642 ± 3029
		95	60	60	60.6 ± 5.1	3.88 ± 0.17	31587 ± 1197
				90	64.4 ± 3.7	3.90 ± 0.02	32661 ± 1163
	85	-	-	-	61.1 ± 2.3	4.41 ± 0.36	34458 ± 433
		75	60	1~7	57.5 ± 3.6	3.82 ± 0.30	29630 ± 1490
				60	57.3 ± 5.9	4.23 ± 0.31	34602 ± 1729
				90	55.9 ± 4.4	4.37 ± 0.22	33925 ± 4651
			80	1~7	44.3 ± 3.0	3.60 ± 0.12	27594 ± 823
				90	48.3 ± 3.9	3.55 ± 0.16	29745 ± 1782
	65	-	-	-	67.1 ± 3.8	4.83 ± 0.48	35039 ± 2877
		65	60	1~7	50.5 ± 3.3	3.53 ± 0.20	30792 ± 2184
				30	57.5 ± 2.7	3.88 ± 0.40	33068 ± 799
				90	57.5 ± 1.7	4.58 ± 0.14	33563 ± 2160
			80	1~7	44.5 ± 7.6	3.42 ± 0.15	27912 ± 1807
				90	52.8 ± 3.5	3.83 ± 0.53	33228 ± 1598
	0	-	-	-	67.6 ± 3.4	4.51 ± 0.45	34916 ± 2015
	vacuum	-	-	-	64.7 ± 6.6	5.37 ± 0.48	36209 ± 2532
		sealed	60	1~7	59.7 ± 1.9	4.35 ± 0.50	31903 ± 431
MLC	**100**	-	-	-	**61.4 ± 1.9**	**4.12 ± 0.24**	**47351 ± 4670**
	95	-	-	-	67.6 ± 9.7	4.23 ± 0.36	34445 ± 1533
	85	-	-	-	78.7 ± 1.8	4.73 ± 0.64	37992 ± 407
		sealed	60	1~7	66.9 ± 4.3	4.06 ± 0.52	35643 ± 685
	65	-	-	-	82.9 ± 4.7	4.64 ± 0.76	38690 ± 828
		sealed	60	1~7	59.6 ± 3.5	3.99 ± 0.03	33178 ± 1394
	0	-	-	-	85.5 ± 1.5	5.28 ± 0.30	39559 ± 2075
	vacuum	-	-	-	81.5 ± 5.1	5.69 ± 0.34	37530 ± 1398
MHC	**100**	-	-	-	**40.6 ± 2.3**	**3.41 ± 0.20**	**35293 ± 4829**
	95	-	-	-	45.2 ± 5	3.80 ± 0.43	26893 ± 3283
	85	-	-	-	52.0 ± 2.5	3.89 ± 0.40	31690 ± 1300
		sealed	60	1~7	42.8 ± 3.5	3.35 ± 0.42	28327 ± 659
	65	-	-	-	53.5 ± 2.1	3.95 ± 0.52	32102 ± 851
		sealed	60	1~7	40.4 ± 2.0	3.77 ± 0.01	28847 ± 862
	0	-	-	-	52.4 ± 4.1	4.70 ± 0.19	31577 ± 2838
	vacuum	-	-	-	55.7 ± 5.1	4.78 ± 0.43	30640 ± 1071

In the following subsections graphical representations of some results from Table 4.2 are presented. Two types of diagrams are displayed for each mechanical property. The first one includes measurements conducted on samples from the concretes MLC, MRC and MHC that were dried at reference temperature. This type of diagram presents the changes in the mechanical properties relative to the reference values as a function of the mean relative humidity of the samples. The mean relative humidities were calculated from measurements of relative humidity according to the procedure described in Appendix B.1. The second type of diagram includes measurements conducted on samples from the concrete MRC that were sealed and then subject to elevated temperatures. This type of diagram presents the changes in the mechanical properties relative to the reference values as a function of temperature. In the legend of the diagram the different layers are named according to the relative humidity that the samples reached after the conditioning at 20 °C before increasing the temperature and the relative humidity at which they were stored is indicated between brackets. Measurements from samples of the concrete MRC drying at elevated temperatures during 30, 60 and 90 days are not presented here, they will be used to validate the material model in Chapter 6. Additional measurements conducted on samples from the concretes MLC and MHC are presented in Appendix B.3.

4.3.1 Compressive strength

Fig. 4.14 presents the results of compressive strength tests conducted on samples from all three concretes mixtures after drying at reference temperature and different ambient relative humidities.

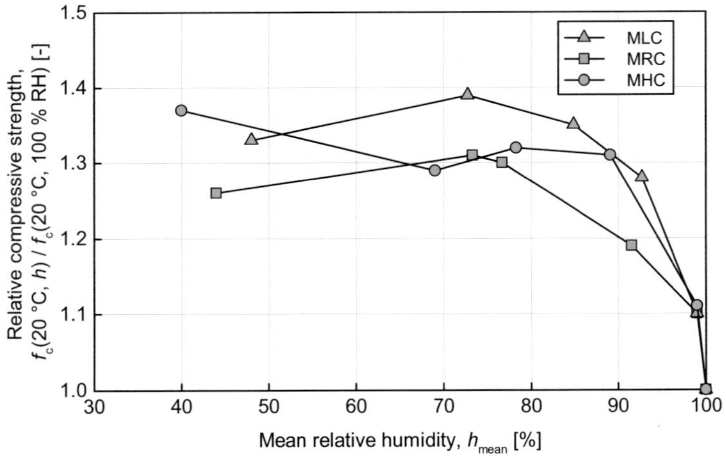

Figure 4.14: Relative compressive strength of the concretes after drying at 20 °C in relation to moisture content at reference conditions

In Fig. 4.14 the calculated value of mean relative humidity for every measured point corresponds to the mean value reached by the samples after being subject to drying according to the drying conditions at 20 °C described in Chapter 3.2.1. All three concretes show a clear tendency of increasing compressive strength with decreasing moisture content within the range of 100 % to 80 % RH, reaching an increment of the compressive strength of around 35 %. By relative humidities below 80 % the enhancement of the compressive strength ceases. The results suggest that a constant value settles after reaching a relative humidity of around 80 %. The results do not indicate any influence of the concrete microstructure. The concretes MLC, MRC and MHC behave similarly. All the measured points can be allocated within a ±5 % stripe.

After drying at reference temperature some samples were sealed and subject to elevated temperatures. The results of compressive strength from such samples belonging to concrete MRC are presented in Fig. 4.15. This diagram shows the influence of temperature on the compressive strength of concrete taking into account the history of moisture content of the concrete before heating. In the legend of the diagram the layers are named according to the mean relative humidities reached by the samples before the heating took place. The corresponding relative humidities at which the samples were conditioned at reference temperature are specified in parentheses.

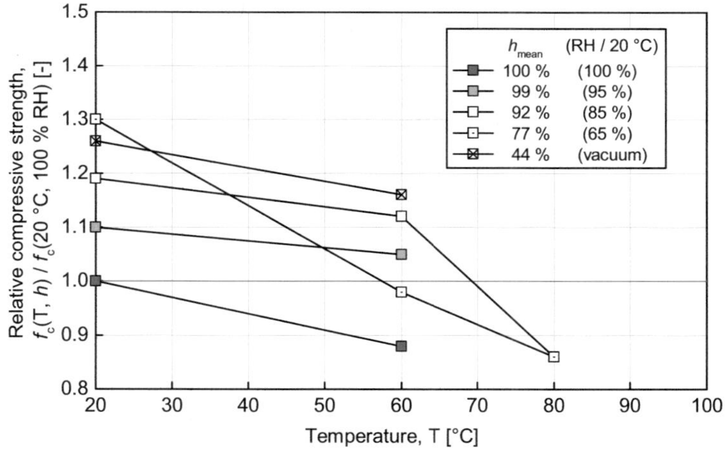

Figure 4.15: Relative compressive strength of concrete MRC heated at different moisture contents

In general the diagram indicates that concrete compressive strength reduces with increasing temperature. However, the negative effect of temperature on the concrete compressive strength varies depending on the moisture content. Assuming that this negative effect is due to incompatibilities between the thermal coefficient of expansion of the aggregates and the hardened cement paste, it is expected that the samples holding the highest amount of water in their microstructure should show the lowest values of compressive strength at higher temperatures. This assumption holds true when comparing the results from the samples having mean relative humidities of 100 %, 77 %, and 44 %. However, the samples with 99 % and 92 % mean relative humidity do not seem to follow the same tendency. In case temperature could solely influence the concrete samples through thermal expansion, the measured points of these two group of samples should lie between the samples heated up with mean relative humidities of 100 and 77 %.

As mentioned in Chapter 4.2, concretes heated while having a high amount of moisture experienced a densification of their microstructure most probably due to hydrothermal reactions between silicone dioxide of the aggregates and calcium hydroxide from the hardened cement paste. The results presented in Fig. 4.15 suggest that hydrothermal reactions could have also influenced the mechanical behaviour of the concrete by healing the damage induced by the thermal expansion incompatibilities, which explains the unexpected behaviour of the samples having a mean relative humidity of 99 % and 92 %. For the case of the samples having a mean relative humidity of 100 %, the influence of hydrothermal reactions is discarded because these samples were heated and tested within a day, which did not allow enough time for the hydrothermal reactions to develop.

4.3.2 Tensile strength

Fig. 4.16 presents the results of tensile strength tests carried out on samples of the three concretes after drying at reference temperature and several environmental relative humidities in relation to the values of tensile strength obtained at reference conditions.

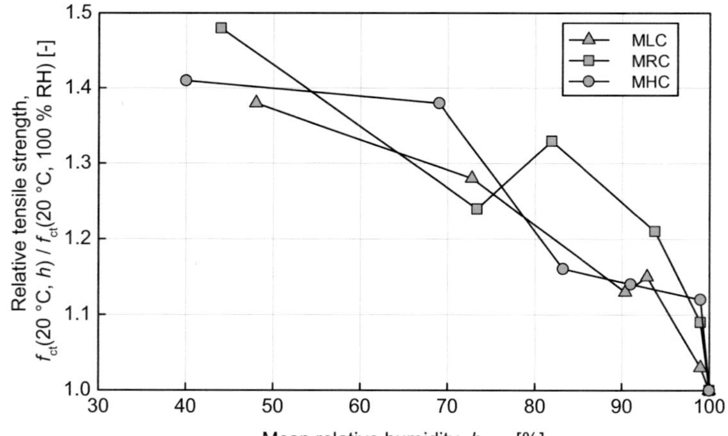

Figure 4.16: Relative tensile strength of the concretes after drying at 20 °C in relation to moisture content at reference conditions

At reference temperature, with decreasing water content the tensile strength of concrete increases. This statement is evidenced by the measurements conducted on concretes which were previously subject to drying at 20 °C. The relative values of tensile strength show an increase of the strength within the whole range of relative humidities that were achieved in the samples. The concretes reached an increment of around 40 % of the tensile strength by a mean internal relative humidity in the range of 40 to 50 %. Unlike the results from compressive strength tests, the tensile strength do not seem to reach a maximum value within the range of relative humidity that was measured. Fig. 4.16 suggests that the tensile strength of the concretes can continue increasing if the concretes reach lower mean values of internal relative humidity. By comparing the three concretes between them, no influence of the concrete microstructure could be identified. All concretes follow the same path and the points lie within a ±5 % stripe.

Temperature affects the concrete tensile strength negatively. Fig 4.17 presents the influence of temperature on the tensile strength of the concrete MRC according to the mean moisture content of the samples before the temperature was increased. In general, independent of the water content, all samples showed a reduction of the tensile strength with an increment of the temperature.

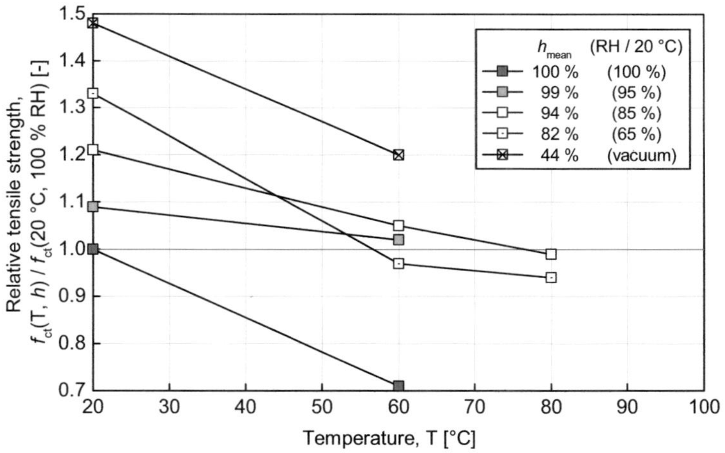

Figure 4.17: Relative tensile strength of concrete MRC heated at different moisture contents

Similar to the measurements of compressive strength, the samples that were maintained up to 7 days at elevated temperature while containing a high amount of moisture (mean relative humidities of 99 % and 94 %) behaved differently. If the influence of temperature on the concrete would be limited to only damage due to the thermal expansion incompatibilities, the values from the samples heated up with mean relative humidities of 99 % and 94 % should lie between the ones from the samples with 100 % and 82 %. Again, this is not the case because of the development of hydrothermal reactions that contributed to a relative increase of the tensile strength by partially healing the damage caused by thermal incompatibilities.

4.3.3 Modulus of elasticity

Contrary to the results from the tests of compressive and tensile strength, after drying, the modulus of elasticity does not reach values higher that the ones measured at reference conditions. Fig. 4.18 shows the influence of drying on the modulus of elasticity of the investigated concretes. The values of relative modulus of elasticity in all concretes with mean relative humidities below 100 % are lower than 1.0, which denotes that an increment of the modulus of elasticity could not be seen. Particularly remarkable is the behaviour of the samples having a mean relative humidity around 99 %. The modulus of elasticity changes abruptly when the concrete dries from a mean relative humidity of 100 % to 99 % reaching the lowest measured values. As drying continues, the modulus of elasticity increases and partially recovers, although the reference value of stiffness was not achieved by the samples dried within the measured range of relative humidities.

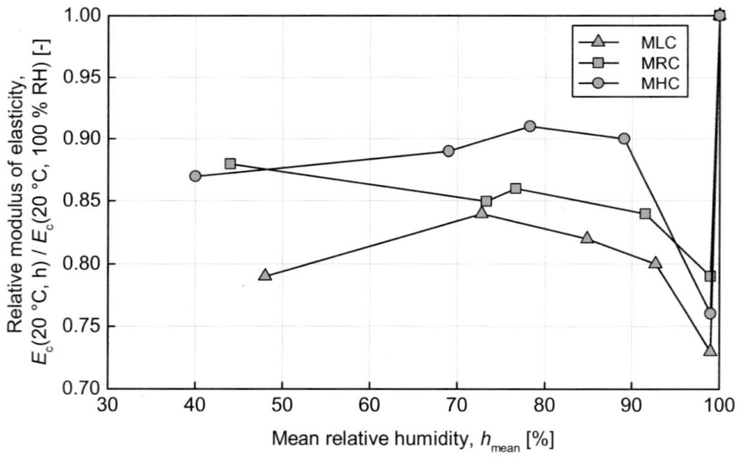

Figure 4.18: Relative modulus of elasticity of the concretes after drying at 20 °C in relation to moisture content at reference conditions

According to Grübl [64], concretes under saturated conditions have a higher modulus of elasticity than dry concretes, because the water in the pores is incompressible and cannot scape so quickly when subject to rapid loading. The results from Fig. 4.18 suggest that the effect of water contribution to the loading capacity vanishes once the water finds a way to scape. It seems that only a small change in the relative humidity of the concrete pores is enough to settle a path for water to leak and therefore an abrupt change in the modulus of elasticity takes place. However, further drying increases the friction between the components of the hardened cement paste microstructure, which leads to a recovery of the modulus of elasticity. According to the measurements, this recovery ceases at a mean relative humidity of around 80 % and is maintained nearly constant down to a relative humidity of 40 to 50 %. Unlike the results from strength tests, a dependency of the concrete microstructure on the modulus of elasticity may be identifiable. After the abrupt change in the modulus of elasticity the recovery is the higher, the higher the porosity (w/c-ratio) of the concrete mixture. Same as by the strength measurements, all the measured points are allocated within a ± 5 % stripe.

Fig. 4.19 illustrates the influence of temperature on the modulus of elasticity of the concrete MRC. In the same way as Figs. 4.15 and 4.17, in the legend the layers are named according to the mean relative humidities reached by the samples before the temperature was raised and the values of stiffness are relative to the modulus of elasticity measured at reference conditions (20 °C / 100 % RH).

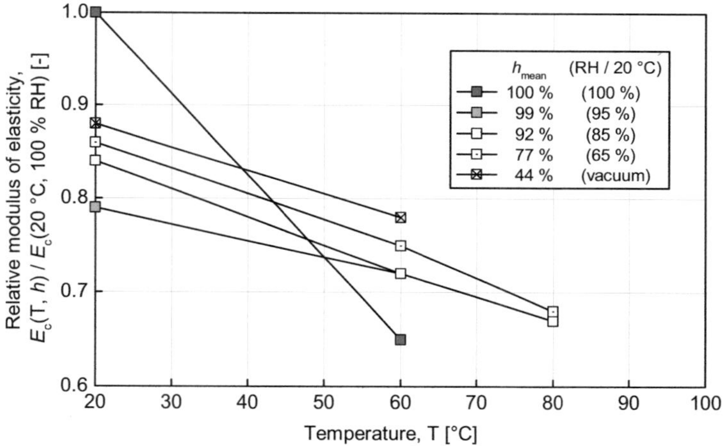

Figure 4.19: Relative modulus of elasticity of concrete MRC heated at different moisture contents

Analogously to the results from tests of compressive and tensile strength, the temperature influences the modulus of elasticity negatively. The results suggest a linear relation between damage and temperature for all the moisture contents. The assumption of higher damage by higher moisture content of the samples is reflected in the results, which implies that the contribution of hydrothermal reactions on the modulus of elasticity is only secondary. This may be due to the fact that by testing modulus of elasticity the samples are loaded with only one third of the compressive strength. At lower stresses the concrete behaves linearly and healing of microcracks do not necessarily make a difference as is the case by stresses approximating failure. The value of modulus of elasticity measured by the samples heated while having 100 % RH in the concrete pores is connected in the diagram with the reference at 20 °C following the same approach for the presentation of the results from Figs. 4.15 and 4.17. At first glance, this measurement seems not to be quite correct and should be located above the measurement of the samples tested with a mean relative humidity of 44 %. However, taking into account the abrupt change in the modulus of elasticity presented by the concretes after a small change in the relative humidity by humidities close to 100 %, this point can only be located above the one from the samples having a mean relative humidity of 44 % if the sealing of the samples was so outstanding that no movement of water due to temperature could have taken place. As this was most probably not the case, the point by 100 % mean relative humidity and 60 °C actually represents samples that suffered an abrupt change in the modulus of elasticity due to a minor drying and in addition the influence of temperature.

4.4 Measurements of concrete creep and shrinkage

It is generally accepted that both creep and shrinkage deformations are composed of a basic and a drying component. The basic component is understood as the deformation which occurs if there is no moisture exchange of a member with the environment. This is the case at very high ambient humidities RH > 98 % or if a member is sealed (e.g. by coating) or for rather thick members where the moisture loss in the near-surface zone may be neglected compared to the total moisture content of the concrete member. The drying component is defined as the additional deformation which may be observed if a moisture exchange with the environment takes place; for strain definitions refer to [N15].

The time dependent behaviour of concrete was investigated by means of testing creep and shrinkage at 20, 40 and 70 °C at a relative humidity of 65 %. Before testing, the samples were previously conditioned for over 450 days at 20 °C and 65 %, 85 % and 95 % RH (see Chapter 3.2.5). The measurements of shrinkage started with the beginning of loading, which together with the fact that the samples were tested at ages over 500 days, implies that the measured deformations correspond solely to drying shrinkage occurring during loading. Basic and drying shrinkages taking place before loading were not captured by the measurements. In the following diagrams the shrinkage deformation is presented as positive, even though it corresponds to a volume reduction, and is plotted in time according to the duration of loading of the creep tests. The results from creep measurements are presented as specific creep which is calculated by dividing the creep deformation by the stress applied on the samples. The creep deformation is calculated by subtracting the elastic deformation measured on the loaded samples during the incrementation of the load until it reached the desired value and the shrinkage deformation measured on the companion shrinkage samples from the total deformation measured on the loaded samples. Presenting the results of the creep tests by means of specific creep allows to compare the results of samples that were subject to different levels of loading.

The following subsections present a selection of the results from the tests of creep and shrinkage. The measurements of creep and shrinkage were conducted at intervals ranging from 2 seconds to 10 minutes. However, for sake of simplicity measurements separated by similar intervals in a logarithmic time scale were selected and are shown in the following figures represented by the symbols of the legend. In addition, lines connecting the symbols are used to facilitate the reading of the diagrams. The results presented here are used to analyse how the concretes behaved according to the influences of previous conditioning, concrete microstructure, and temperature. The test results, including all the conducted measurements and information about conditioning and loading regimes of the creep tests, are presented graphically in Appendix B.4.

4.4.1 Influence of water content

Due to the different conditions at which the concrete samples were subject previous to testing, the samples had different water contents at the moment of loading. Both creep and shrinkage are the higher, the higher the water content of the concrete and, more specially, the faster the concrete losses water with the environment [98]. It is therefore expected that for the same concrete, the samples containing the highest amount of water, at the beginning of the test, exhibit the maximum shrinkage and creep deformations. Fig. 4.20 shows the development in time of the shrinkage deformations measured on the concrete MRC on tests conducted at reference temperature. In the legend of the diagram the layers are named according to the conditioning of the samples before the measurements were conducted.

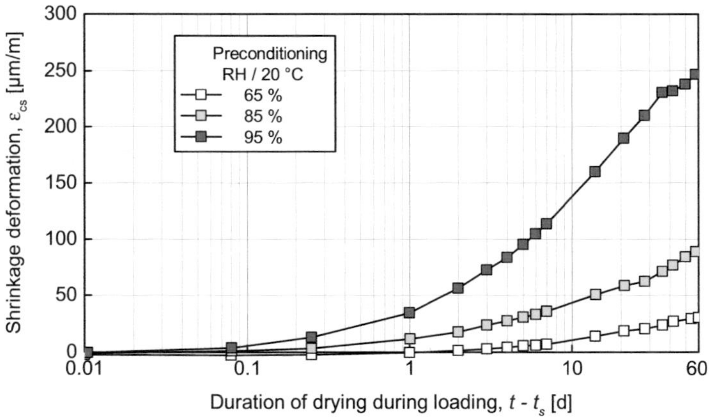

Figure 4.20: Time development of the shrinkage deformation of the concrete MRC tested at 20 °C and 65 % RH according to the conditioning prior to test

The samples previously conditioned at 65 % RH shrink very little while the samples conditioned at higher relative humidities show significant higher shrinkage deformations, especially those conditioned at 95 % RH. According to the theory, the measurements of creep deformations shall also show the same tendency. Fig. 4.21 presents the evolution in time of the specific creep from the creep samples of concrete MRC tested at 20 °C.

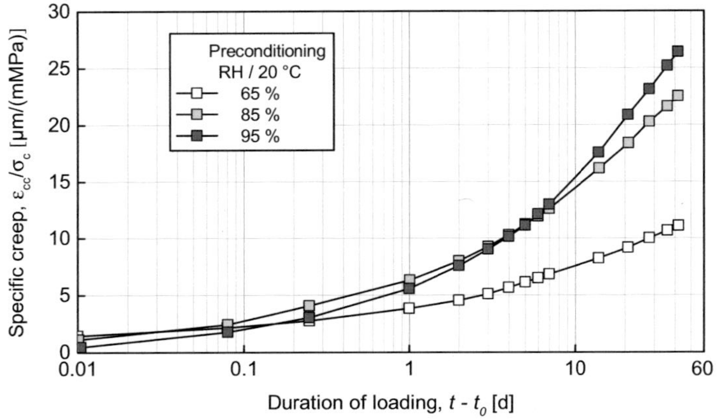

Figure 4.21: Time development of the specific creep of the concrete MRC tested at 20 °C and 65 % RH according to the conditioning prior to test

Again, the creep samples follow the expected tendency. The same behaviour is also seen in the tests conducted on the concretes MLC and MHC at reference temperature, as well as on all three concretes at elevated temperatures, both for creep and shrinkage deformations (see Appendix B.4.5 to B.4.8).

4.4.2 Influence of concrete microstructure

The concrete mixtures that were investigated differ in the quantity of water used, which directly influences the microstructure generated within the hardened cement paste. The lower the w/c-ratio, the lower the porosity, and therefore, the higher the strength of the concrete samples. Regarding creep and shrinkage, lower porosity, i.e. low w/c-ratio and high degree of hydration, reduces the amount of deformation [98]. On this background, it is expected that the concrete with the lowest w/c-ratio (MLC) shows the lowest creep and shrinkage deformations and consequently, the concrete with the highest w/c-ratio (MHC) the highest. The following diagrams compare the shrinkage and creep deformations of the concretes MLC, MRC and MHC on the basis of a common conditioning scheme before as well as during the tests. Fig. 4.22 presents the time development of the shrinkage deformations of the three concrete mixtures tested at reference temperature and previously conditioned at 20 °C and 85 % RH.

In Fig. 4.22, the expected dependency of the shrinkage deformation on the concrete porosity cannot be clearly seen. The sample from the concrete MLC behaved a bit uncommon, and even though by the last measured values its shrinkage deformation lies below the deformation of the concrete MRC, the measurements do not provide any conclusive results. The concretes behaved unexpectedly which could question the validity of the measured values. Consequently, these measured values can be taken as consistent, if the unexpected behaviour is also seen in the measurements of the creep samples.

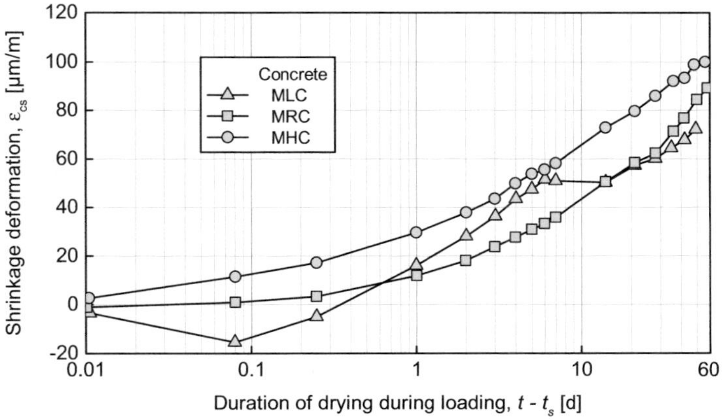

Figure 4.22: Time development of the shrinkage deformation of the concretes MLC, MRC and MHC tested at 20 °C / 65 % RH and previously conditioned at 20 °C / 85 % RH

Fig. 4.23 presents the time development of the specific creep from samples of the concretes tested at reference temperature and previously conditioned at 20 °C and 85 % RH.

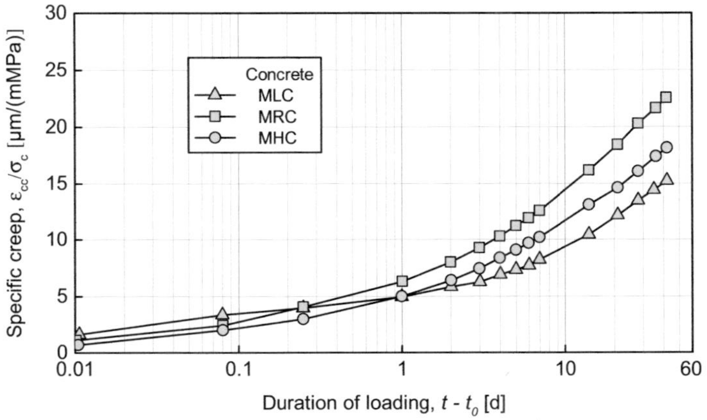

Figure 4.23: Time development of the specific creep of the concretes MLC, MRC and MHC tested at 20 °C / 65 % RH previously conditioned at 20 °C / 85 % RH

As well as by the shrinkage measurements, when comparing the three concrete mixtures in Fig. 4.23, no dependency of the specific creep on the concrete microstructure can be identified. Similar behaviour is seen when the mixtures are compared with each other on the basis of a common conditioning procedure for samples tested at elevated temperatures, both for shrinkage as well as for creep (see Appendix B.4.5 to B.4.8). Based on these results, it can be stated that, for the procedure of conditioning and testing followed for the samples, no influence of the concrete microstructure could be recognized. The fact that concretes with lower porosity develop lower creep and shrinkage deformations has already been studied and verified in the past through a large number of experimental investigations. The results presented in this investigation shall not be used to controvert that evidence anyhow. Therefore, at this point the approach is oriented to finding an explanation that can disclose the logic of such an unexpected behaviour.

The reason why the measured creep and shrinkage deformations do not agree with the fact that they should be the lower, the lower the porosity of the samples, can be provided by analysing the conditioning of the samples previous to testing. In Figs. 4.22 and 4.23 the three concretes are compared with each other on the basis of the previous conditioning at which they were subject. Even though they all were conditioned in the same chambers during barely the same time before the creep tests were carried out, due to the differences in their concrete microstructures, their water contents at the moment of testing were not necessarily equivalent. As it was already discussed in Chapter 4.1 (see Figs. 4.3 and 4.7), the lower the porosity, the slower the drying process of the concrete. Hence, after the same conditioning procedure, samples from concrete MRC had more water in their microstructure than samples from concrete MHC and less than those from concrete MLC. Consequently, being the creep and shrinkage deformations so dependent on the water content and water loss during loading (see Chapter 4.4.1), the way the samples were conditioned previous to testing does not apply as reference to compare the deformation behaviour of the concrete mixtures with each other because each concrete mixture reached a different moisture content at the moment of testing. This can only be possible if the concretes are compared on the basis of a mean relative humidity of 100 % in the pore system, which was not a case of study in this investigation.

4.4.3 Influence of concrete temperature

Temperature has a significant impact on the development of creep and shrinkage deformations mainly because of the influence that it exerts on the water transport process within the concrete microstructure. According to Müller and Kvitsel [98] creep and shrinkage deformations increase when the concrete is tested at elevated temperatures, while subjecting the concrete to long-acting elevated temperature before testing leads to a decrease of the deformations. Regarding the tests conducted in the investigation, the load was applied one hour after the temperature within the climate chamber was reached, which cannot be considered as a long-acting elevated temperature. Therefore, it is expected that the measured shrinkage and creep deformations are the higher, the higher the temperature during test. Fig. 4.24 shows the development of the shrinkage deformations of the concrete MRC tested at 20, 40 and 70 °C after being conditioned at 20 °C and 65 % RH.

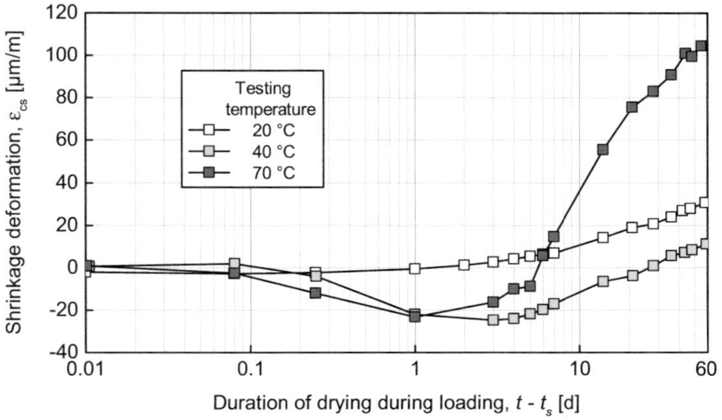

Figure 4.24: Time development of the shrinkage deformation of the concrete MRC tested at 65 % RH and 20, 40 and 70 °C and previously conditioned at 20 °C / 65 % RH

The measured deformations of the concrete samples tested at 40 and 70 °C decreased up to approximately one day after the beginning of the test. These negative values were caused by thermal expansion suffered by the samples after the loads were set. As mentioned in Chapter 3.2.5, the mean temperature reached by the concrete samples at the moment of loading was around 2 to 3 °C below the desired temperature. After imposing the load, it took around 24 hours for the samples to reach the temperature of the chamber. On account of these small temperature changes within the first day, the measurements of deformation at the beginning of the tests actually correspond to a superposition of shrinkage and thermal expansion. During the first 24 hours, the thermal expansion was larger than the shrinkage and therefore the measured deformations were negative.

In Fig. 4.24, the results of the test conducted at 70 °C complies with the expected behaviour. The shrinkage deformation is evidently higher when testing at 70 °C compared to 40 or 20 °C. At first sight, the samples tested at 20 °C seem to develop higher shrinkage deformation than those tested at 40 °C. However, taking into account the influence of thermal expansion within the first day of measurements, this may not be the case. After the samples tested at 40 °C stop expanding, i.e. the lowest point in the diagram is reached, the time development of the shrinkage deformation shows a steeper slope than the one from the samples tested at 20 °C, which suggests that the actual shrinkage deformation of the samples tested at 40 °C is higher.

The measurements of creep deformation should also comply with the expected behaviour of higher deformation at higher testing temperatures. Fig. 4.25 shows the development of the specific creep of samples from the concrete MRC previously conditioned by 20 °C and 65 % RH and then tested at 20, 40 and 70 °C by a relative humidity of 65 %.

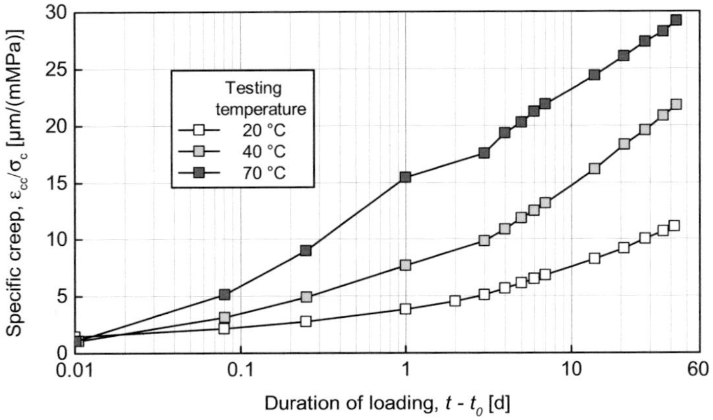

Figure 4.25: Time development of the specific creep of the concrete MRC tested at 65 % RH and 20, 40 and 70 °C and previously conditioned at 20 °C / 65 % RH

In general, at higher testing temperatures, higher creep deformations are developed. The results presented in Fig. 4.25 as well as further results included in the Appendix B.4.7 and B.4.8 reflect this statement very well. Nevertheless, it is mandatory to examine whether the fact that at the moment of loading the desired testing temperature was not entirely reached by the samples could have influenced not only the shrinkage deformation but also the resulting creep deformations. As it was already mentioned in Chapter 3.2.5, the heating regime influences the test results significantly. According to investigations conducted by Budelmann [35], samples subject to loading during heating show larger creep deformations than samples loaded after reaching the temperature of testing. During the first day after loading, the samples suffered a small temperature increment that may have affected the development of the creep deformations. However, the measurements do not show any indications of uncommon development of the creep deformation within the first 24 hours of loading. Therefore, the conducted tests can be assumed as valid to investigate the development of creep on samples loaded at elevated temperatures.

4.5 Summary and conclusions

The chapter presents the most relevant results of the experimental investigation. The conducted experiments underline the importance of water content and its interaction with temperature on the mechanical behaviour of concrete. In general, the behaviours shown by the concrete samples agreed with the assumptions made by the design of the experimental program. The measurements of relative humidity showed that the drying process of concrete accelerates with the elevation of temperatures and is the slower, the lower the porosity of the concrete microstructure. Additionally, the measurements confirmed the increase of the relative humidity on the concrete pores due to an increment of the temperature as denoted by the hygrothermic coefficient of concrete.

The results from the experiments of mercury intrusion porosimetry and gas permeability were consistent and agreed very well with each other. They suggest that at elevated temperature and high moisture content a development of hydrothermal reactions between the silicone dioxide from the aggregates and calcium hydroxide from the cement paste caused densification of the microstructure in the concrete samples. Regarding the tests to evaluate the mechanical properties, the results reflect the negative influence of temperature and the positive effect of drying on the tensile and compressive strength as well as on the modulus of elasticity except for the samples tested at 20 °C and 100 % RH which show the highest modulus of elasticity. Creep and shrinkage deformations also agree with the expectation of being the higher, the higher the temperature and moisture content of the concrete by testing.

Besides delivering results that are consistent and in line with the expected behaviours, the experimental investigation also clarified the following issues that light the way for the formulation of the new concrete material model.

- Concretes drying at higher relative humidities do not reach the equilibrium with the environment faster than those drying at lower ambient relative humidities, which denotes that the diffusion coefficient of concrete cannot depend solely on the absolute content of moisture of the concrete.
- Elevated temperatures influence concrete in such a way that the differences seen in the water transport properties between concretes with low porosity (w/c = 0.4) and high porosity (w/c = 0.6) tested at elevated temperature are less significant than those seen in concretes tested at room temperature.
- The assumption that the influence of temperature on the mechanical properties of concrete can be explained by superposition of the effects caused by temperature and moisture content does not apply to concretes with high moisture content. A third effect known as hydrothermal reactions emerges from combining high temperature and high moisture content for concretes made with siliceous aggregates withdrawing the possibility of linear superposition.
- The influence of drying on concrete compressive and tensile strength is not dependent on the microstructure of the concrete. The measurements of modulus of elasticity showed a dependency on the concrete microstructure, however, differences between the measured points of the mixtures can be allocated within a ±5 % stripe and therefore, these differences can be neglected.

- Elevated temperatures influence the strength and stiffness of concrete negatively. Any recovery of these mechanical properties with time must be accompanied by a loss of moisture or associated with further hydration or the development of hydrothermal reactions.
- In order to compare the creep and shrinkage behaviour of concretes with different microstructures, the initial condition of moisture in the samples have to be the same. Due to the strong influences that water content exerts on the creep and shrinkage deformations and the fact that the drying process of the samples vary according to the porosity of their microstructure, no valid comparisons between different concrete mixtures can be carried out on the basis of a common drying procedure. Therefore, the results from creep and shrinkage tests could not deliver any conclusive information about the influence of the concrete microstructure.

The tendencies shown by the results from the extensive experimental investigation follow the expected behaviour according to the considerations presented in Chapter 2.3.2. Only the assumption taken about the hydrothermal reactions can be considered as inaccurate. By comparing the strength of samples subject to elevated temperatures and high environmental relative humidity, Budelmann [35] found no significant differences between samples exposed to these conditions during 20 and 90 days. On this basis, it was assumed that the influence of hydrothermal reactions on the strength and stiffness of the samples can be neglected. However, the results obtained in this study showed that subjecting the samples to conditions of elevated temperatures and high environmental relative humidity up to 7 days influences the tensile and compressive strength of the samples considerably. Therefore, the influences of hydrothermal reactions on the strength of concrete are not insignificant, they just occur faster than expected, probably in a period of time shorter than a week.

Through a meticulous analysis of the results, the factors that play a main role in the impact of temperature on the mechanical properties of concrete were disclosed. Hence, the experimental results can be used to formulate models to predict the mechanical behaviour of concrete with the advantage of precisely knowing the physical background of this behaviour.

Chapter 5

Predicting the mechanical behaviour of concrete

The literature evaluated in Chapter 2 together with the results from the conducted experimental investigation discussed in Chapter 4 made clear that the effects of elevated temperatures below 100 °C on the concrete mechanical properties are strongly linked to the water contained in the concrete at the moment of heating as well as during the length of the heating period. Hence, knowing the relation between temperature exposure and moisture content is mandatory to accomplish the objective of formulating models to predict the mechanical behaviour of concrete subject to elevated temperatures.

By means of the experimental program, the relationships between temperature and moisture content and their influence on the mechanical properties of concrete were identified. Based on the experimental results, in the following sections, mathematical expressions are formulated to describe the behaviour shown by the concrete samples according to the measurements of pore relative humidity and temperature of exposure.

5.1 Strength and stiffness

Following the assumptions made for the conception of the experimental program (see Chapter 2.3.2), the strength and stiffness of concrete subject to elevated temperatures can be described by superposition of the effects of drying and thermal incompatibilities if the effects of acceleration of hydration and hydrothermal reactions are neglected. On this background, Eq. 5.1 to 5.3 calculate the influence of temperature and moisture content on a given mechanical property β by summing the effects of drying D_β and thermal incompatibility S_β and multiplying them with the property at reference conditions $\beta(h_{\mathrm{ref}} = 100\ \%,\ T_{\mathrm{ref}} = 20\ °\mathrm{C})$.

$$f_{\mathrm{c}}(h,T) = f_{\mathrm{c}}(h_{\mathrm{ref}},T_{\mathrm{ref}}) \cdot \left(1 + D_{f_{\mathrm{c}}} + S_{f_{\mathrm{c}}}\right) \tag{5.1}$$

$$f_{\mathrm{ct}}(h,T) = f_{\mathrm{ct}}(h_{\mathrm{ref}},T_{\mathrm{ref}}) \cdot \left(1 + D_{f_{\mathrm{ct}}} + S_{f_{\mathrm{ct}}}\right) \tag{5.2}$$

$$E_{\mathrm{c}}(h,T) = E_{\mathrm{c}}(h_{\mathrm{ref}},T_{\mathrm{ref}}) \cdot \left(1 + D_{E_{\mathrm{c}}} + S_{E_{\mathrm{c}}}\right) \tag{5.3}$$

The effects of drying and thermal incompatibility are dependent on the moisture content of the concrete which can be account for by means of the relative humidity of the concrete pores. However, as it was shown in Chapter 4.1.3, the relative humidity in the concrete pores is dependent on the temperature. Hence, in order to make the equations consistent, the moisture content is defined as the relative humidity of the concrete pores at reference temperature $h_{T_{\text{ref}}}$. This relative humidity corresponds to the relative humidity that the concrete pores with the same moisture content would have at reference temperature ($T_{\text{ref}} = 20$ °C) and can be calculated using the definition of the hygrothermic coefficient of concrete $K(h_{T_{\text{ref}}})$ described in Chapter 2.1.2.2 as follows:

$$h_{T_{\text{ref}}} = h_T - K(h_{T_{\text{ref}}}) \cdot (T - T_{\text{ref}}) \tag{5.4}$$

where h_T is the relative humidity of the concrete pores at a given temperature T[°C]. The unknown parameters of Eqs. 5.1 to 5.3 can be obtained from the experimental results by defining functions that fit the measurements adequately. The effects of drying can be addressed by considering the tests conducted at reference temperature (T_{ref} = 20 °C) and the effects of thermal incompatibilities arise from the tests conducted at elevated temperatures once the effects of drying are excerpted from the results.

As mentioned in Chapter 4.3, due to the fact that the relative humidity of the tested samples varied along the radius, the test results were presented using mean values of the relative humidity h_{mean}, which were calculated following the procedure described in Appendix B.1. However, in the following mathematical formulations $h_{T_{\text{ref}}}$ instead of h_{mean} is used because the relative humidity $h_{T_{\text{ref}}}$ does not necessarily has to be a mean value of the concrete sample owing that the effects of moisture content on the mechanical properties of concrete shall vary with the radius in the same way the relative humidity does (e.g. in case of a pointwise calculation). Therefore a differentiation is done by relating the following diagrams, showing experimental results, to h_{mean} and the formulations, in a more general way, to $h_{T_{\text{ref}}}$. The relative humidity $h_{T_{\text{ref}}}$ is dimensionless, i.e. it reaches values between 0 and 1.0, and the temperature is given in [°C].

5.1.1 Effects of drying

The strength of concrete increases with decreasing water content in the concrete pores. Eq. 5.5 and 5.6 describe the development of the factors D_{f_c} and $D_{f_{ct}}$, accounting for the influence of drying on the compressive and tensile strength of concrete, according to the relative humidity of the concrete pores at reference temperature $h_{T_{\text{ref}}}$.

$$D_{f_c} = -0.32 \cdot (h_{T_{\text{ref}}})^{26} + 0.32 \tag{5.5}$$

$$D_{f_{ct}} = -0.44 \cdot (h_{T_{\text{ref}}})^{4} + 0.44 \tag{5.6}$$

Fig. 5.1 presents the results of the tests of compressive and tensile strength conducted on the concretes MLC, MRC and MHC at 20 °C and different mean relative humidities. The results are presented relative to the reference values in the same way as in Chapter 4. Based on Eqs. 5.5 and 5.6, the maximal increase of the compressive strength due to drying is 32 % and is reached at a relative humidity of the concrete pores around 82 % while the tensile strength increases up to 44 % if the concrete dries down to around 30 % RH. Although the scatter of the results is considerably high, the majority of the measured points are located within a 5 % distance from the calculated curves.

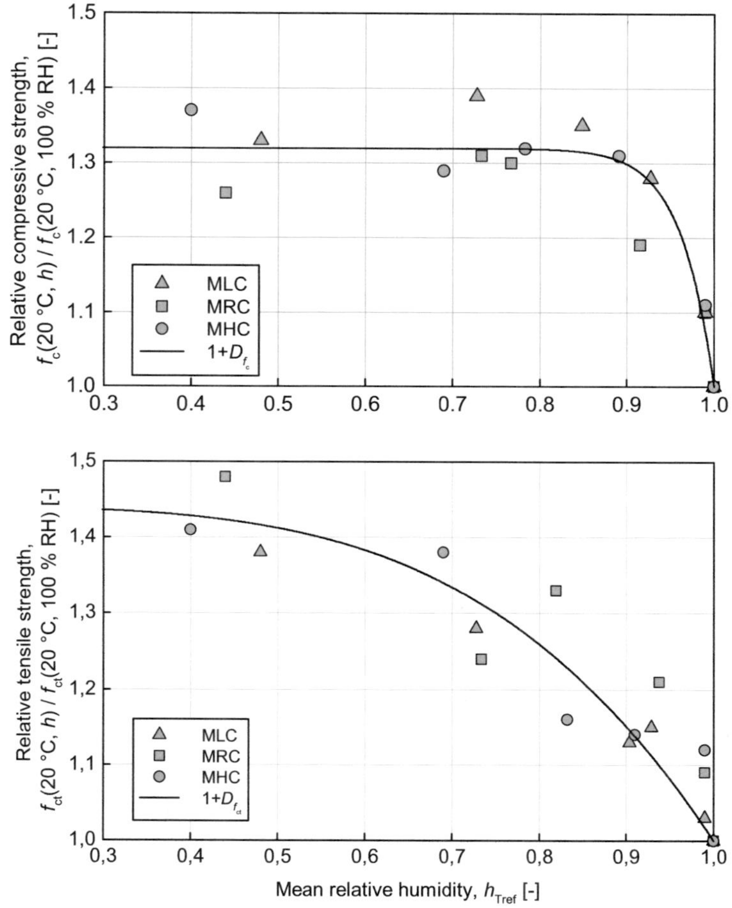

Figure 5.1: Relative compressive strength (top) and relative tensile strength (bottom) of the concretes in relation to moisture content at reference temperature. Comparison between calculated values and measurements

The influence of drying on the stiffness of concrete substantially differs from the influence on the concrete strength. The values of modulus of elasticity reached by the concrete after drying are lower than those shown at reference conditions of moisture. As it was discussed in Chapter 4.3.3, at a relative humidity in the concrete pores of 100 %, water contributes to the loading capacity of the concrete. However, at lower relative humidities the water finds a path to leak and this contribution vanishes, leading to an abrupt change in the modulus of elasticity at relative humidities closely below 100 %. This effect can be mathematically described by a piecewise function as follows:

$$\begin{aligned} D_{E_c} &= 0.0 && \text{for } h_{T_{ref}} = 1.0 \\ D_{E_c} &= -0.12 \cdot (h_{T_{ref}})^{19} - 0.14 && \text{for } h_{T_{ref}} < 1.0 \end{aligned} \tag{5.7}$$

Fig. 5.2 presents the results of the tests of modulus of elasticity conducted on the concretes MLC, MRC and MHC at 20 °C and different mean relative humidities. According to Eq. 5.7, the modulus of elasticity sinks to around 74 % of its original value at reference conditions after a small change in the relative humidity takes place and then, as the concrete dries further, the modulus of elasticity partly recovers. The recovery of the modulus of elasticity finalizes at a relative humidity of around 78 % where it reaches its maximum corresponding to 86 % of the stiffness that the concrete had at reference conditions of humidity and temperature.

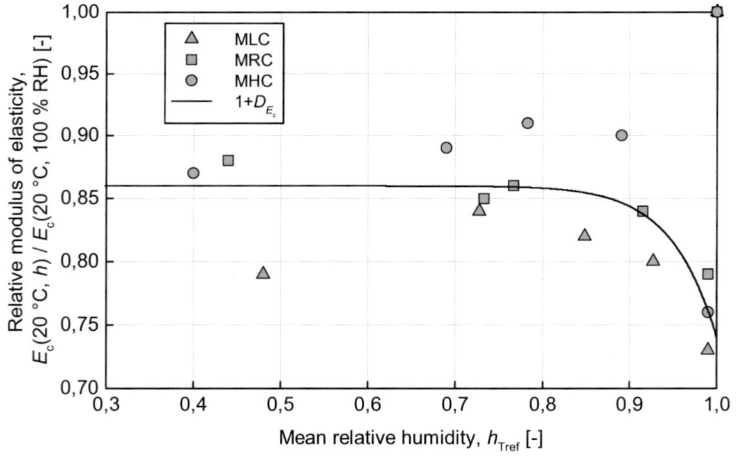

Figure 5.2: Relative modulus of elasticity of the concretes in relation to moisture content at reference temperature. Comparison between calculated values and measurements

Even though the measured results show a possible dependency of the recovery of the modulus of elasticity on the w/c-ratio, the scatter of the results between all three concretes is similar to those seen in the tests of strength. Therefore, using an unique equation to describe the recovery of the modulus of elasticity at relative humidities below 100 % for all three concretes seems to be reasonable.

5.1.2 Combined effects of temperature and moisture content

The samples tested at elevated temperatures after being subject to drying for a long period of time, were influenced by the effects of drying and thermal incompatibility. In addition to these two effects, the samples heated while containing a high amount of water, which correspond to those conditioned at 20 °C and 95 % RH as well as 85 % RH, were affected by hydrothermal reactions as denoted in Chapter 4.2. The effects of hydrothermal reactions were identified only qualitatively and therefore it is not possible to include them in the formulations. This could only be achieved if the time-development function of the effects that the hydrothermal reactions caused on the mechanical properties of concrete were known. Hence, for the formulation of the following equations, only the experimental results from the samples previously conditioned or tested in a manner that the influence of hydrothermal reactions can be neglected, are employed (see Chapter 4.3.1).

Eqs. 5.8 to 5.11 correspond to the mathematical approximation of the influence of temperature on the concrete compressive and tensile strength considering the effect of thermal incompatibility. Following the measurements, the concrete gains in strength during the drying process at 20 °C. Thereafter the temperature is increased causing a reduction of the concrete strength that is dependent on the water content of the concrete at the moment of heating. Hence, these equations are dependent on the temperature and the relative humidity of the concrete pores at temperature of reference $h_{T_{\mathrm{ref}}}$.

$$S_{f_{\mathrm{c}}} = 0.35 \cdot \left(\frac{T-20}{100}\right)^2 + a_{f_{\mathrm{c}}} \cdot \left(\frac{T-20}{100}\right) \tag{5.8}$$

$$a_{f_{\mathrm{c}}} = 2.3 \cdot 10^{-9} \cdot \exp\left(19.1 \cdot h_{T_{\mathrm{ref}}}\right) - 0.55 \cdot \exp\left(0.64 \cdot h_{T_{\mathrm{ref}}}\right) \tag{5.9}$$

$$S_{f_{\mathrm{ct}}} = 0.45 \cdot \left(\frac{T-20}{100}\right)^2 + a_{f_{\mathrm{ct}}} \cdot \left(\frac{T-20}{100}\right) \tag{5.10}$$

$$a_{f_{ct}} = 7.6 \cdot 10^{-5} \cdot \exp\left(8.5 \cdot h_{T_{\mathrm{ref}}}\right) - 0.72 \cdot \exp\left(0.48 \cdot h_{T_{\mathrm{ref}}}\right) \tag{5.11}$$

Fig. 5.3 compares the calculations with the values of the experimental measurements of compressive and tensile strength conducted on the concrete MRC at different temperatures and moisture contents relative to the values at reference conditions. The calculations represented by the continuous lines include the effects of both drying and thermal incompatibility. The starting points of the curves at a temperature of 20 °C is given by the equations that describe the effects of drying (Eqs. 5.5 and 5.6), and the reduction of the strength with increasing temperature is ruled by the equations that account for the effect of thermal incompatibility (Eqs. 5.8 to 5.11).

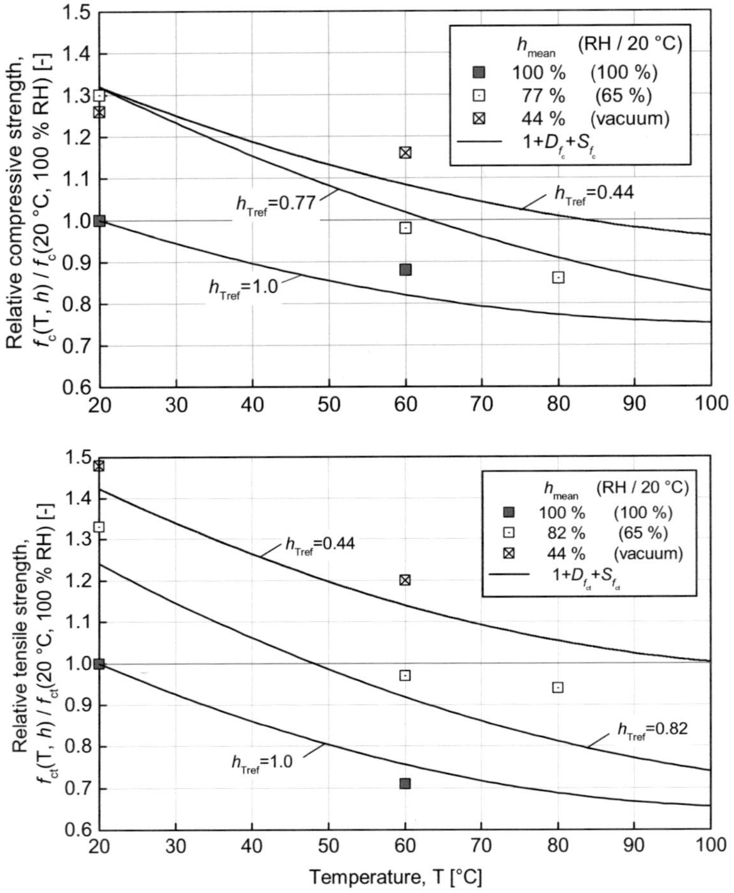

Figure 5.3: Relative compressive strength (top) and tensile strength (bottom) of concrete MRC heated at different moisture contents. Comparison between calculated values and measurements

Following the experimental results, after increasing strength due to drying at 20 °C, increasing the temperature reduces the strength. The experiments conducted at 80 °C suggest that the relation between strength reduction and temperature is not linear and therefore polynomials of second order were proposed for Eqs. 5.8 and 5.10. These equations are purely approximations of the mechanical behaviour shown by the samples. They look somewhat complicated because they are defined for a large range of temperatures and relative humidities ($0.3 \leq h_{T_{ref}} \leq 1$, $20\ °C \leq T \leq 100\ °C$). Due to the scatter of the results, it cannot be expected that the calculations follow the measured points precisely. However, the curves describe the behaviour of the concrete strength appropriately.

Following the same approach used for the formulation of the equations to describe the influence of thermal incompatibility on the strength of concrete, Eqs. 5.12 and 5.13 account for the influence of thermal incompatibility on the modulus of elasticity. Differing from the formulations presented to describe the influence of thermal incompatibility on the strength of concrete in which second order polynomial were selected, according to the measurements, a linear relation between temperature and the reduction of the modulus of elasticity can be assumed (see Chapter 4.3.3).

$$S_{E_c} = a_{E_c} \cdot \left(\frac{T-20}{100} \right) \tag{5.12}$$

$$a_{E_c} = 2.3 \cdot 10^{-5} \cdot \exp\left(9.0 \cdot h_{T_{ref}}\right) - 0.2 \cdot \exp\left(0.7 \cdot h_{T_{ref}}\right) \tag{5.13}$$

The comparisons of the calculations of the modulus of elasticity with the results of the experiments conducted on the concrete MRC are presented in Fig. 5.4.

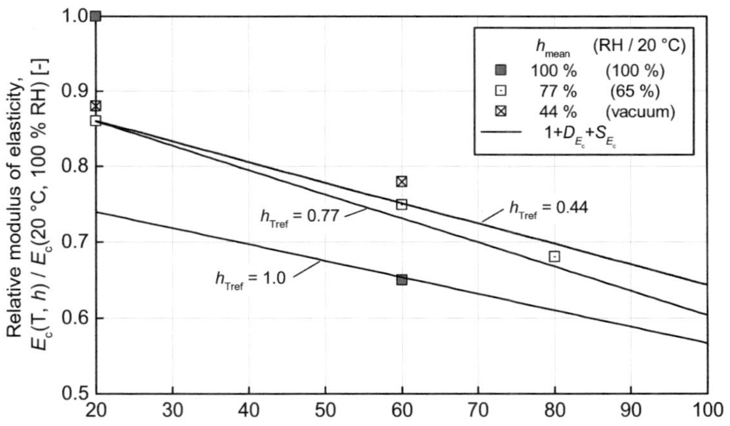

Figure 5.4: Relative modulus of elasticity of concrete MRC heated at different moisture contents. Comparison between calculated values and measurements

The behaviour of concrete at different temperatures and pore relative humidities can be represented in a 3D diagram as surfaces. The curves presented in Figs. 5.3 and 5.4 correspond to the intersection of these surfaces with planes fixed at a given relative humidity. Diagrams representing the surfaces can be seen in Appendix D.2. The formulations presented above are valid for a range of temperatures between 20 and 100 °C and relative humidities between 0 % and 100 % RH.

5.2 Time dependent deformations

Given the important role that the water content plays in the development of concrete deformations with and without the action of external forces, it shall be meaningful to include the water content in the prediction of the time dependent deformations of concrete. As it was mentioned in Chapter 3.2.2, also during the conduction of the creep and shrinkage tests, the relative humidity of the concrete samples was measured. In the following subsections formulations to describe the strain behaviour of concrete subject to drying and loading at reference temperature as well as at elevated temperatures are developed based on the measurements of deformation and relative humidity. Due to the fact that the relative humidity is non-uniform across a section, both creep and shrinkage deformations are not uniform either. However, the stiffness of the material contributes with the generation of internal stresses which ensure the development of plane deformations across the section. The calibration of the following equations was carried out based on the measurements of plane creep and shrinkage deformations that may be related to the mean relative humidity of the concrete pores h_{mean} calculated from the measurements of relative humidity conducted during the creep tests (see Appendix B.4.1 and B.4.2). Nevertheless, the formulations to calculate creep and shrinkage can be more general if the relative humidity of the concrete pores h is used, which allows to calculate not only plane but also non-uniform shrinkage and creep deformations across the section.

5.2.1 Reference temperature

Any model used to describe the strain behaviour at elevated temperatures must be able to properly predict the behaviour at 20 °C. The measurements conducted at reference temperature help to set the basis of the models. The models are firstly calibrated for the temperature of reference where no influence of temperature is considered and then they can be upgraded to include temperatures above the reference.

5.2.1.1 Basic shrinkage

It was assumed in Chapter 4.4 that the conducted measurements on shrinkage correspond to drying shrinkage due to the advanced age of the concrete samples at the beginning of the measurements. Therefore, is not possible to develop any type of mathematical formulation for the description of basic shrinkage based on the conducted experiments. However, it is well known that basic shrinkage is mainly influenced by the cement composition and the mix design and is independent of the geometry of the concrete member because it occurs even if no moisture loss is possible [98]. In case including formulations for the calculation of basic shrinkage is necessary, they can be adopted from the model of Model Code 2010 [N15] which adequately take into account these characteristics. These formulation are not included here because no measurements of basic shrinkage are available and their accuracy could not be verified.

5.2.1.2 Drying shrinkage

Due to the fact that drying shrinkage obeys the process of drying of the concrete microstructure, a first and reasonable approximation would be to directly relate the changes in the drying shrinkage deformations with changes in the relative humidity of the concrete pores over the whole cross-section. Following this approach, Fig. 5.5 presents measurements from the concrete MRC previously conditioned at 20 °C and 95 % RH and subject to drying at 20 °C and 65 % RH. The vertical axis on the left denotes the shrinkage deformations and the vertical axis on the right the mean relative humidity of the concrete pores. The horizontal axis gives the duration of drying during loading which corresponds to the duration of drying of the shrinkage samples. In Appendix C.2 additional diagrams, similar to Fig. 5.5, show the correlation between shrinkage deformation and mean relative humidity for the concretes MLC, MRC and MHC measured during the conducted tests.

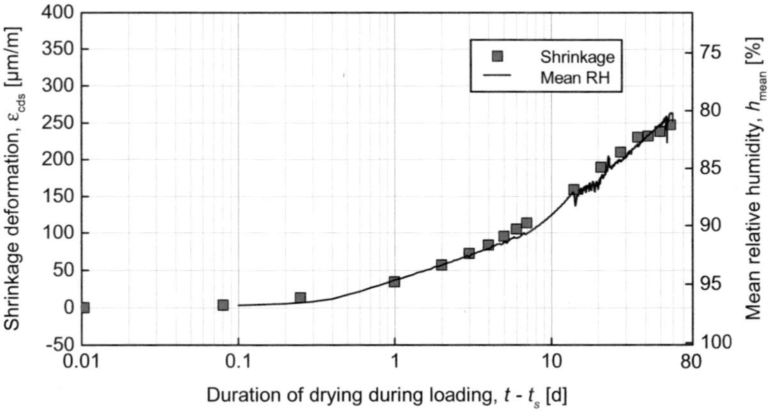

Figure 5.5: Shrinkage deformation and mean relative humidity of the samples from the concrete MRC previously conditioned at 20 °C and 95 % RH during the creep tests at 20 °C and 65 % RH

As it can be seen in Fig. 5.5, the measurements of shrinkage and mean relative humidity correlate very good with each other. Hence, knowing the time-development of the mean relative humidity of the concrete pores of a given concrete geometry, the drying shrinkage can be accurately estimated by introducing a factor of correlation. This can be formulated as follows:

$$d\varepsilon_{\text{cds}} = K_{\text{cds}} \cdot dh \tag{5.14}$$

where K_{cds} corresponds to the factor of correlation between the changes in the drying shrinkage $d\varepsilon_{\text{cds}}$ and the changes in the mean relative humidity dh. The values that K_{cds}

can adopt are dependent on the concrete microstructure. Table 5.1 presents the determined values of K_{cds} based on the conducted experiments.

Table 5.1: Factors to relate changes in shrinkage deformation with mean relative humidity of the concrete samples at 20 °C

Concrete Mixture	w/c-ratio [-]	K_{cds} [μm/m]
MLC	0.40	1650
MRC	0.50	1600
MHC	0.60	1400

According to Table 5.1, the dependency of the factor K_{cds} on the w/c-ratio of the concrete mixtures denotes that, given a change in the relative humidity of the concrete pores, the induced drying shrinkage is the higher, the lower the w/c-ratio of the concrete mixture. This dependency is expressed mathematically in Eq. 5.15.

$$K_{cds}(w/c) = \frac{d\varepsilon_{cds}}{dh} = 1650 \cdot \left[0.78 + \frac{0.22}{1+(1.76 \cdot w/c)^{14}}\right] \tag{5.15}$$

Eq. 5.15 is based on the values of K_{cds} experimentally determined for the three concrete mixtures. The equation describes a s-shaped function which allows to extend the validity of the model to w/c-ratios below and above the ones considered in the measurements. At low values of w/c-ratio the equation reaches a maximum of 1650 and at high values of w/c-ratio it approaches a minimum of 1287.

5.2.1.3 Creep

The creep of concrete is commonly described by means of the creep coefficient φ, which equals the quotient between the creep deformation ε_{cc} and the elastic deformation ε_{ci}. As it was mentioned in Chapter 4.4 the creep deformations from the experimental results were calculated by subtracting the elastic deformations measured on the loaded samples, during the period in which the loads were incremented until they reached the desired value, and the total deformations of the companion shrinkage samples from the total deformations measured on the loaded samples. However, in case of a theoretical approach it is necessary to estimate the elastic deformation ε_{ci} by means of the Hooke's law based on the modulus of elasticity. Furthermore, the modulus of elasticity as well as the compressive strength can vary depending on the moisture content and the temperature of the concrete member as it was assessed experimentally in Chapter 4 and theoretically in previous sections of the present chapter. Therefore, in order to simplify the formulations to calculate creep and make them independent of the history of development of the modulus of elasticity and the compressive strength, the following equations were calibrated using reference values for the modulus of elasticity and the compressive strength for each concrete mixture corresponding to those measured at reference conditions; 20 °C and 100 % RH (see Table 4.2 in Chapter 4).

In a general manner the total creep coefficient can be calculated as the sum of its basic φ_{bc} and drying φ_{dc} components as follows:

$$\varphi(t) = \varphi_{bc}(t) + \varphi_{dc}(t) \tag{5.16}$$

Code type models to calculate creep like those presented in Model Code 2010 [N15] or the second part of the Eurocode 2 [N16] can be used as basis to proposed formulations to calculate the basic and drying creep coefficients. According to Model Code 2010, the basic creep coefficient may be calculated as the product between a factor dependent on the concrete compressive strength $\beta_{bc0}(f_{cm})$ and the time-development function of the basic creep $\beta_{bc}(t,t_0)$ which, in turn, depends on the age of concrete at loading t_0. It follows that:

$$\varphi_{bc}(t,t_0) = \beta_{bc0}(f_{cm}) \cdot \beta_{bc}(t,t_0) \tag{5.17}$$

with:

$$\beta_{bc0}(f_{cm}) = \frac{1.8}{(f_{cm})^{0.7}} \quad \text{and} \quad \beta_{bc}(t,t_0) = \ln\left[\left(\frac{30}{t_{0,\mathrm{adj}}} + 0.035\right)^2 \cdot (t-t_0) + 1\right]$$

where f_{cm} is given in [N/mm²] and $t - t_0$ in [d]. This formulation can be generalized by adding two factors. K_{bc1} to tune the factor $\beta_{bc0}(f_{cm})$ and K_{bc2} included in the time-development function. Eq. 5.17 can then be expressed as:

$$\varphi_{bc}(t,t_0) = \frac{K_{\mathrm{bc1}}}{(f_{cm})^{0.7}} \cdot \ln\left[\frac{t-t_0}{K_{\mathrm{bc2}}} + 1\right] \tag{5.18}$$

The basic creep coefficient can be also formulated incrementally by calculating the derivative of Eq. 5.18 with respect to time, which gives:

$$d\varphi_{bc}(t,t_0) = \frac{K_{\mathrm{bc1}}}{(f_{cm})^{0.7}} \cdot \frac{1}{(t-t_0) + K_{\mathrm{bc2}}} \cdot dt \tag{5.19}$$

In the formulation from Model Code 2010, the factor K_{bc1} is constant and K_{bc2} is a function of the adjusted age at loading $t_{0,\mathrm{adj}}$. Including the age at loading allows the model to predict basic creep on early age concretes. In case of advanced age concretes, the age at loading accounts not only for the state of development of the hydration reactions but also, in an indirect way, for the ageing due to drying. The creep experiments of the present investigation were conducted on concrete samples older than 500 days, and therefore it can be assumed that the hydration process of the concrete samples was completed at the moment of loading. Hence, in regards to the conducted experiments, the age at loading does not represent a relevant parameter. However, given that the samples were subject to different drying scenarios before testing, the moisture content should be important and may be taken into account to define the factors K_{bc1} and K_{bc2}.

Following the approach of the second part of Eurocode 2 [N16], the drying creep coefficient can be calculated as the product of a global parameter φ_{dc0} and the drying shrinkage occurring during loading.

$$\varphi_{dc}(t,t_0) = \varphi_{dc0} \cdot [\varepsilon_{cds}(t) - \varepsilon_{cds}(t_0)] \tag{5.20}$$

with:

$$\varphi_{dc0} = \begin{cases} 1000 & \text{for silica-fume concrete} \\ 3200 & \text{for non silica-fume concrete} \end{cases}$$

In order to keep the same notation used for basic creep, the parameter φ_{dc0} is here replaced by the factor K_{dc} and Eq. 5.20 is rewritten in differential form as:

$$d\varphi_{dc}(t,t_0) = K_{\mathrm{dc}} \cdot d\varepsilon_{cds} \tag{5.21}$$

Incorporating the increment of the drying shrinkage deformation $d\varepsilon_{cds}$ as calculated in Eq. 5.14, Eq. 5.21 becomes:

$$d\varphi_{dc}(t,t_0) = K_{\mathrm{dc}} \cdot K_{\mathrm{cds}} \cdot dh \tag{5.22}$$

The calculation of the drying shrinkage coefficient according to Eq. 5.22 implies that the drying shrinkage can be related to the changes in the moisture content of the concrete, represented by the changes of the relative humidity of the concrete pores dh.

Eqs. 5.19 and 5.22 constitute the main formulations for the creep model at reference temperature. Based on the conducted experiments, the factors K_{bc1} and K_{bc2} from Eq. 5.19 and the factor K_{dc} from Eq. 5.22 can be defined and calibrated. The inputs for the calibration are the creep strains calculated for the samples, the time-dependent mean relative humidity of the samples calculated from the measurements of relative humidity conducted during the loading period and the mechanical and physical properties of the different concrete mixtures at reference conditions.

At reference temperature as well as at elevated temperatures, all creep tests were conducted at an ambient relative humidity of 65 %. Therefore, the samples previously conditioned at 20 °C and 65 % RH exchanged only a small amount of moisture with the environment during the creep tests. This implies that almost no drying shrinkage took place during the tests and consequently the creep measured corresponds, almost entirely, to basic creep. On this background, the creep measurements from the samples previously conditioned at 20 °C and 65 % RH were used to calibrate the factors $K_{\mathrm{bc}1}$ and $K_{\mathrm{bc}2}$ which led to the following equations:

$$\mathrm{K}_{bc1} = 1.55 \cdot \left[0.6 + \frac{0.4}{1+(2 \cdot w/c)^{10}}\right] \cdot h \tag{5.23}$$

$$\mathrm{K}_{bc2} = 0.2 \cdot h \tag{5.24}$$

Both factors K_{bc1} and K_{bc2} depend on the relative humidity of the concrete pores h. In addition, $K_{\mathrm{bc}1}$ depends on the concrete microstructure described by a s-shaped function regulated by the w/c-ratio of the concrete mixture. Knowing the factors for the definition of the basic creep coefficient, the calibration of the factor K_{dc} can be carried out based on the tests conducted on the samples previously conditioned at relative humidities higher than 65 %, namely 85 % and 95 %. Table 5.2 presents the calibrated values of K_{dc}.

Table 5.2: Factors to relate changes in drying creep deformation with shrinkage deformation of the concrete samples at 20 °C

Concrete Mixture	*w/c-ratio* [-]	K_{dc} [µm/m]
MLC	0.40	3400
MRC	0.50	3300
MHC	0.60	3100

Based on the values from Table 5.2, at reference temperature K_{dc} is only dependent on the concrete microstructure and can be described by the following s-shaped function.

$$K_{\mathrm{dc}}(T_{\mathrm{ref}}, w/c) = \frac{d\varphi_{dc}}{d\varepsilon_{cds}} = 3415 \cdot \left[0.88 + \frac{0.12}{1+(1.85 \cdot w/c)^{11}}\right] \tag{5.25}$$

Eqs. 5.19 and 5.22 can be solved if the time-development of the relative humidity h in the concrete pores is known. Using the mean relative humidity of the samples h_{mean}, calculated from the measurements of relative humidity conducted during the creep tests (see Appendix B.4.1), the time-development of the plane creep deformation was calculated. Fig. 5.6 presents the comparison between measured values of creep and calculations.

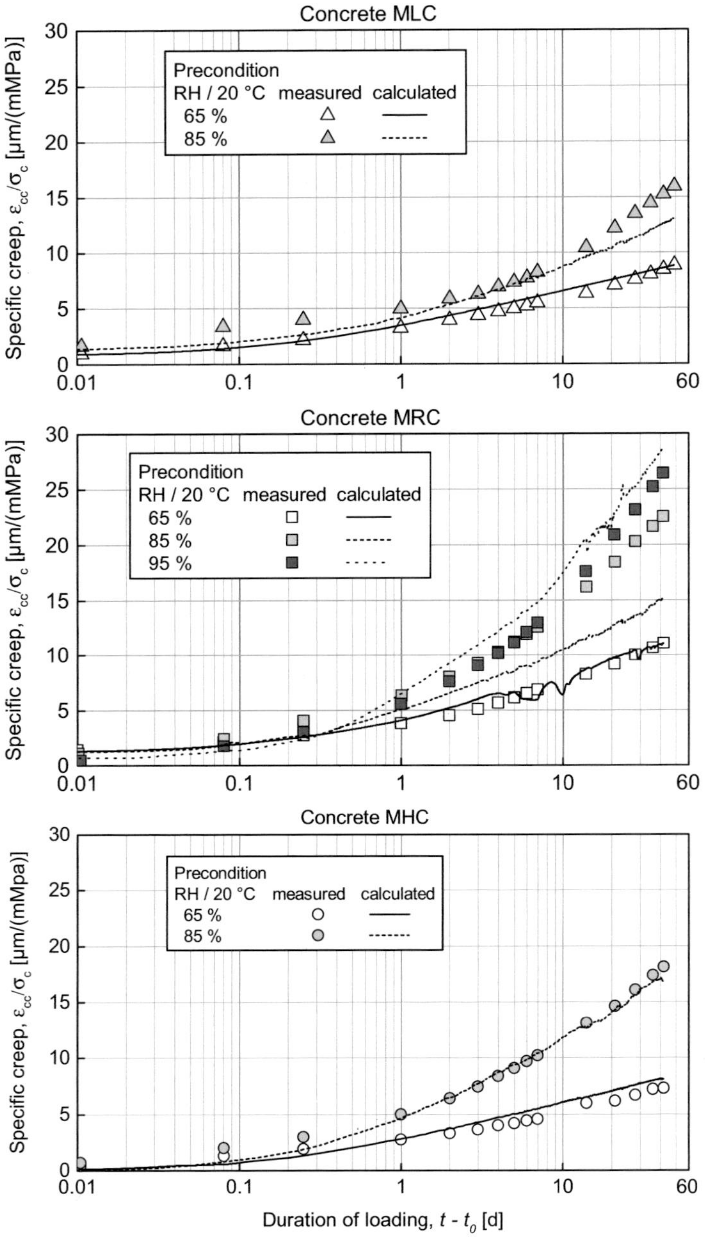

Figure 5.6: Specific creep calculated with the mean relative humidity measured in the concretes in comparison with the measured values for the concretes MLC, MRC and MHC during the creep tests conducted at 20 °C and 65 % RH

For the calculation of the basic creep coefficients, according to Eq. 5.19, the mean compressive strength of the concretes f_{cm} needs to be considered. Moreover, the specific creep corresponds to the quotient between the creep strain ε_{cc} and the applied stress σ_c which can also be calculated as the quotient between the creep coefficient φ and the modulus of elasticity E_c. The values of mean compressive strength and modulus of elasticity used for the calculation of the curves presented in Fig. 5.6 correspond to those measured in the concrete samples at reference conditions (see Table 4.2 in Chapter 4). The calculated curves presented in Fig. 5.6 follow the measured values very well. Only a considerable underestimation can be seen by the calculations for the concrete MRC previously condition at 20 °C and 85 % RH. Provided that the time-development of the relative humidity of the concrete pores is known, the equations can estimate the development in time of creep accurately.

5.2.2 Elevated temperatures

During the experiments of creep and shrinkage conducted at elevated temperatures, the relative humidity in the concrete samples was also measured (see Appendix B.4.2). Based on the knowledge acquired in these investigations, the formulations presented before can be improved in order to include the influence of elevated temperatures. In the following segments additional factors that account for the influence of temperature are presented to complement the prediction of shrinkage and creep based on the time-development of the relative humidity in the concrete pores h. These formulations can be assumed to be valid for a temperature range between 20 °C and approx. 100 °C.

5.2.2.1 Basic shrinkage

Similar to the experiments conducted at 20 °C, the shrinkage measurements at elevated temperatures were conducted on advanced age concrete samples for which no further basic shrinkage was expected to occur. It can be assumed that elevated temperatures influence basic shrinkage by accelerating the process without affecting the final value of deformation. A logical approach to include these effects can be taken from [N15].

5.2.2.2 Drying shrinkage

As presented before, variations of the mean relative humidity of the concrete pores can be related to changes in the drying shrinkage of concrete by multiplying them by the factor K_{cds} described in Eq. 5.15. This factor is dependent on the w/c-ratio of the concrete mixture and can be considered as independent of temperature. In order to account for the influence of temperature, Eq. 5.14 is multiplied by the factor $K_{\text{cds}T}$ and becomes:

$$d\varepsilon_{\text{cds}} = K_{\text{cds}T} \cdot K_{\text{cds}} \cdot dh \tag{5.26}$$

The factor $K_{\text{cds}T}$ was calibrated according to the measurements conducted at temperatures higher than the reference of 20 °C. The calibration led to the following equation.

$$K_{\text{cds}T} = 1.2 - 0.01 \cdot T \tag{5.27}$$

According to Eq. 5.27, with increasing temperature, the factor relating changes in the relative humidity of the concrete pores with drying shrinkage deformation decreases.

In Fig. 5.7 the shrinkage deformations measured on companion specimens during the loading period of the creep tests conducted on the concrete MRC and MHC at 70 °C and 65 % RH are plotted with the calculations carried out based on Eq. 5.26.

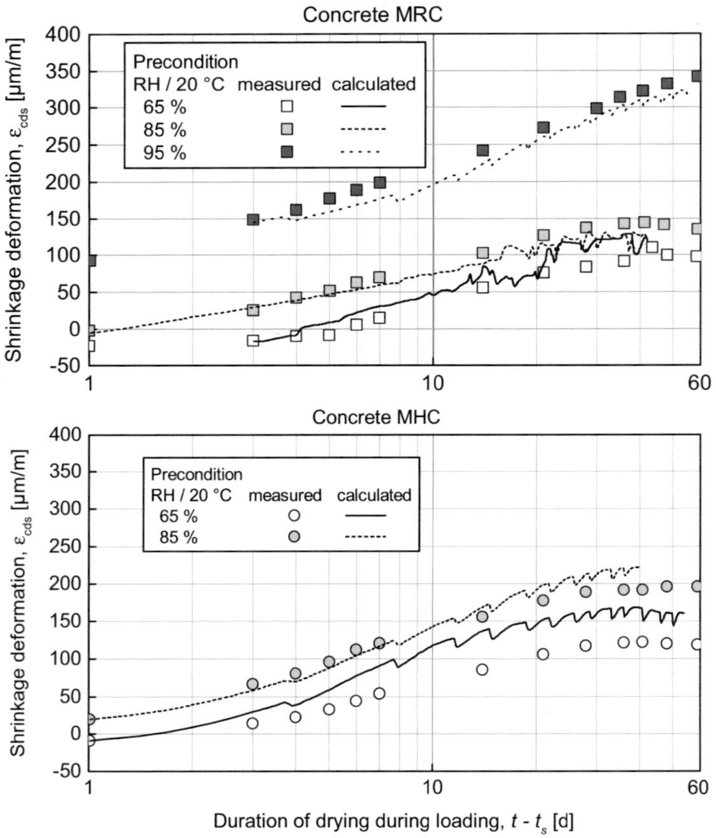

Figure 5.7: Drying shrinkage calculated with the mean relative humidity measured in the concrete pores in comparison with the measured values for the concrete MRC (top) and MHC (bottom) on companion specimens during the creep tests conducted at 70 °C and 65 % RH

The developed model of shrinkage at elevated temperatures can also follow the measured data very accurately. The calculated curves in Fig. 5.7 start at later points in time (1 to 3 days) than the measurements of shrinkage because, for the calculations, the measurements of relative humidity are only useful if the temperature within the concrete samples has already stabilised. During the incrementation of the temperature, the relative humidity of the concrete pores changes but these changes shall not be related with shrinkage deformations. As it was discussed by the analysis of the measurements of shrinkage at elevated temperatures in Chapter 4.4.3, at the moment of application of the load, the temperature of the samples was around 2 to 3 °C below the desired test temperature and it took around 24 h for the samples to reach the temperature of the chamber and even longer (up to 3 d) for the relative humidity of the concrete pores to reach a maximum value and start decreasing.

In addition to the comparisons from Fig. 5.7, similar calculations, providing also satisfactory results, were carried out for the concrete MLC at a testing temperature of 70 °C and the concrete MRC tested at 40 °C (see Appendix C.2).

5.2.2.3 Creep

Following the formulations presented before, creep at 20 °C can be modelled based on the development of the mean relative humidity of the concrete pores during the loading period. The relative humidity in the concrete pores is influenced by temperature, and therefore basic creep and drying creep as described by Eqs. 5.19 and 5.22 shall be influenced by temperature too. Hence, new factors $K_{\mathrm{bc}T}$ and $K_{\mathrm{cds}T}$ need to be consider in these equations in order to account for the influence of temperature leading to the following equations:

$$d\varphi_{bc}(t,t_0) = K_{\mathrm{bc}T} \cdot \frac{K_{\mathrm{bc}1}}{(f_{cm})^{0.7}} \cdot \frac{1}{(t-t_0)+K_{\mathrm{bc}2}} \cdot dt \tag{5.28}$$

$$d\varphi_{dc}(t,t_0) = K_{\mathrm{cds}T} \cdot K_{\mathrm{cds}} \cdot K_{\mathrm{dc}} \cdot dh \tag{5.29}$$

Eq. 5.28 results from multiplying Eq. 5.19 by the factor $K_{\mathrm{bc}T}$ to account for the influence of temperature on the development of basic creep. In case of drying creep, temperature affects the development of drying shrinkage and consequently also drying creep, as assumed in Eq. 5.21. Therefore, the influence of temperature on drying creep is already covered if the influence of temperature on drying shrinkage is considered by multiplying Eq. 5.22 by the factor $K_{\mathrm{cds}T}$ as presented in Eq. 5.29.

The factor $K_{\mathrm{bc}T}$ was calibrated based on the conducted experiments. The calculations performed with the values presented in Table 5.3 showed good agreement with the measured data.

Table 5.3: Factor accounting for the influence of temperature on basic creep

Concrete Mixture	w/c-ratio [-]	K_{bcT} 20 °C	40 °C	70 °C
MLC	0.40	1.0	-	3.1
MRC	0.50	1.0	2.5	3.0
MHC	0.60	1.0	-	2.5

The calibration of the factor K_{bcT} suggests that it is dependent on the w/c-ratio and the temperature. The dependency of K_{bcT} on the w/c-ratio can be expressed by a s-shaped function and the dependency of temperature is covered by including, within the function, two temperature-dependent parameters as proposed in Eq. 5.30.

$$K_{\mathrm{bcT}} = a_{K_{\mathrm{bcT}}} \cdot \left[b_{K_{\mathrm{bcT}}} + \frac{1 - b_{K_{\mathrm{bcT}}}}{1 + (2.2 \cdot w/c)^{-16}} \right] \tag{5.30}$$

The temperature-dependent parameters $a_{K_{\mathrm{bcT}}}$ and $b_{K_{\mathrm{bcT}}}$ are equal to 1.0 at 20 °C and vary with temperature according to Eqs. 5.31 and 5.32:

$$a_{K_{\mathrm{bcT}}} = 1 + 2.7 \left(\frac{T - 20}{100} \right)^{0.35} \tag{5.31}$$

$$b_{K_{\mathrm{bcT}}} = 1.088 - 0.0044 \cdot T \tag{5.32}$$

where T is given in [°C]. Analysing the factor K_{bcT} and K_{cdsT} help to understand the behaviour of creep at elevated temperatures in comparison with 20 °C. The factor influencing basic creep K_{bcT} increases with increasing temperature, leading to basic creep deformations three times higher at 70 °C than at 20 °C (see Table 5.3). On the other hand, the factor influencing drying creep K_{cdsT} decreases 50 % at 70 °C (see Eq. 5.27). This means that for the same change on the relative humidity dh the drying creep deformations of concrete at elevated temperatures are lower than at 20 °C. This observation can be explained by analysing the storage capacity of the concrete. The higher the temperature, the lower the amount of water, that the concrete can store (see Chapter 2.1.2.2) which implies that changes on the relative humidity of the concrete pores induce less moisture loss at elevated temperatures than at 20 °C and therefore lower drying creep deformations are seen. Fig. 5.8 shows results from the creep tests conducted at 70 °C in comparison with the calculations based on Eqs. 5.28 and 5.29.

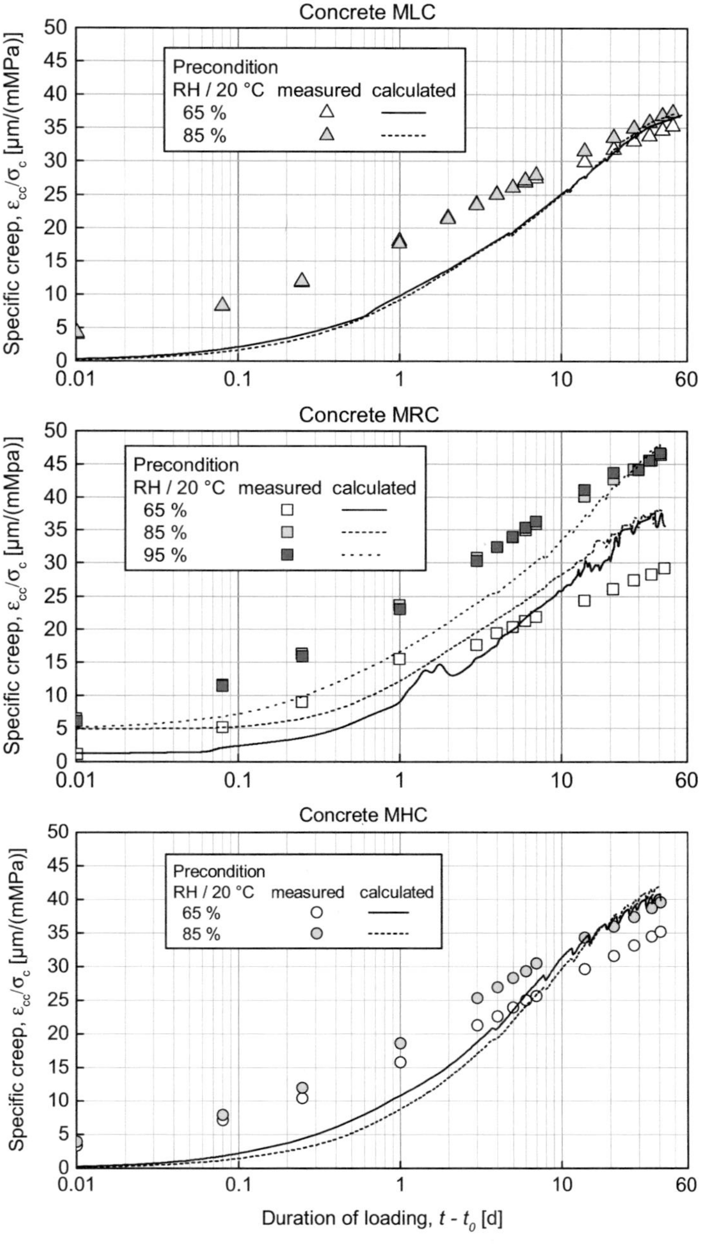

Figure 5.8: Specific creep calculated with the mean relative humidity measured in the concrete pores in comparison with the measured values for the concretes MLC, MRC and MHC during the creep tests conducted at 70 °C and 65 % RH

As new dependencies are added to the formulations, the accuracy of the results is affected. Nevertheless, as it can be seen in Fig. 5.8, the calculations achieved a good correlation with the measured data. Further comparisons with the results of the tests conducted at 40 °C for the concrete MRC, also showing acceptable correlations, are presented in Appendix C.3.

In conclusion, basic creep increases with increasing temperature while drying creep, although develops faster because of the acceleration of the drying process, reaches lower final values at elevated temperatures in comparison to 20 °C. These observations are in accordance with the results from Schwesinger et al. and Seki and Kawasumi [137, 139]. They found that the quotient between creep deformations at elevated temperatures and creep deformations at 20 °C is higher by sealed samples than by unsealed samples. This can be understood as follows: in case of sealed samples, only basic creep develops while by unsealed samples basic and drying creep take place. The quotient is therefore lower by unsealed samples because the drying creep component decreases with increasing temperature.

5.3 Summary and conclusions

Formulations were developed with the objective of calculating the mechanical behaviour of concrete subject to diverse conditions of temperature and relative humidity. Based on the conducted experiments, basic equations to describe the influence of temperature and drying on the compressive strength, tensile strength and the modulus of elasticity of concrete were formulated. Approaches to calculate creep and shrinkage were also presented using the conducted measurements of relative humidity as reference.

The calibration of the formulations to describe the influence of temperature and drying on the mechanical properties of concrete brought new knowledge and contributed with the comprehension of the physical background of the problem. The most important conclusions derived from the present chapter are summarized in the following items.

- The dependency of the strength and stiffness of concrete on the conditions of temperature and moisture content can be described mathematically through simple equations under the assumption that these conditions influence the mechanical properties independently.
- The rate of change of the mean relative humidity within concrete samples is found to scale in a unique manner the rate of drying shrinkage, independent of the moisture content. This behaviour is valid for concretes tested at 20 °C as well as at elevated temperatures.

- Due to the fact that basic creep is influenced by the moisture content of the concrete, the equations formulated suggest that the drying creep, drying shrinkage and the mean relative humidity follow the same kinetics.
- The total creep deformation increases with increasing temperature. While the component of basic creep is the higher, the higher the temperature, the component of drying creep reaches, although faster, lower final values at elevated temperatures than at 20 °C.

Chapter 6

Development of a new material model for concrete

Besides the gain in knowledge about the physical background behind the mechanical properties of concrete, the conducted experiments were aimed at developing a material models to predict the behaviour of these mechanical properties at elevated temperatures below 100 °C. As it was first addressed in Chapter 2 and later observed in Chapters 4 and 5, the mechanical properties of concrete are highly dependent on the moisture content of the material. Elevated temperatures not only affect the concrete mechanical properties directly, they also facilitate the movement of water within the concrete pores causing further changes to the mechanical properties due to changes in the water content of the concrete microstructure. In consequence, the formulations to predict the mechanical properties of concrete from Chapter 5 were based on the measurements of relative humidity of the concrete pores.

Being the moisture content of concrete so relevant for its mechanical properties, the effects of temperature can only be accurately modelled if the time-dependent process of moisture transport is taken into account. This chapter presents the development of a time and temperature-dependent model of moisture transport in concrete. Subsequently, a new material model capable of predicting the influence of variable conditions of temperature and moisture content on the mechanical properties of concrete is presented. This material model is composed by the model of water transport coupled with the formulations to assess compressive strength, tensile strength, modulus of elasticity, shrinkage and creep of concrete at elevated temperatures presented in Chapter 5.

6.1 Modelling the transport of moisture in concrete

Following the model from Bažant and Najjar [11] presented in Chapter 2.2.3, the drying of concrete can be described as a diffusion process where the relative humidity in the pore cavities changes with time towards reaching equilibrium with the relative humidity of the environment and can be calculated by means of the Fick's second law. This approach requires the solution of a non-linear differential equation which can be easily achieved using numerical methods, like, for instance, finite differences. The difficulty lies in requiring the use of special measuring devices to record the relative humidity in the concrete pores during a long period of time. Once the time-dependent distribution of the relative humidity in the concrete samples is known, the model can be calibrated. Nevertheless, the universality of the models based on such measurements is very poor. So far the available experiments in the literature are limited and in most of the cases the calibrations have been carried out for a specific condition of temperature and environmental relative humidity and therefore the model is only valid for a single condition specified by the experiments. On this background, in the experimental investigations presented in Chapter 4, water content (mass of water per volume of concrete) and relative humidity in the pore cavities were measured for concretes with w/c-ratios between 0.4 and 0.6 drying in steady conditions at 85 %, 75 % and 65 % RH at temperatures ranging from 20 °C up to 80 °C to create a sufficiently broad data basis. These measurements allowed to gather enough information to calibrate a more general model capable of calculating the moisture transport for environmental conditions beyond those used in the experiments.

In order to develop the model of water transport in concrete, two main equations need to be formulated. The first equation corresponds to the diffusion equation given in Chapter 2.2.3 (see Eq. 2.4) which describes the water transport at constant temperature and the second equation corresponds to the hygrothermic coefficient of concrete which accounts for the changes in the relative humidity of the concrete pore cavities due to a temperature change (see Eq. 2.3 in Chapter 2.1.2.2). Taking advantage of the axis-symmetry of the concrete samples and the fact that they were sealed on the upper and bottom faces, it can be assumed that the moisture content within the samples varies only in the direction of the radius. Therefore, the solution of the moisture transport equation can be simplified by solving only the radial dimension in polar coordinates. The main equations of the model are then expressed as follows:

$$\frac{\partial h(r,t)}{\partial t} = \frac{1}{r} \cdot \frac{\partial}{\partial r} \left(D(h,T) \cdot r \cdot \frac{\partial h(r,t)}{\partial r} \right) \tag{6.1}$$

$$K = \left(\frac{dh}{dT} \right)_{u=const} \tag{6.2}$$

where h is the relative humidity in the concrete pores, r is the radial position, t is time, D is the diffusion coefficient, T is the temperature and K is the hygrothermic coefficient of concrete. This coefficient takes into account the variation of the relative humidity dh in de concrete pores due to a temperature change dT at a given water content u.

Eq. 6.1 has no analytical solution, however, it can be solved numerically. The solution of the equations was implemented in a MATLAB script by means of an implicit finite difference method using backward difference for the time derivative and a first and second order central difference for the space derivatives. The MATLAB script for the case of a cylindrical sample subject to a constant relative humidity of the environment at constant temperature is presented in Appendix F.1.

According to the model from Bažant and Najjar, the diffusion coefficient D is calculated as the product of the diffusion coefficient at reference conditions D_1 (i.e. $h = 1.0$ and $T = 20$ °C) and two functions accounting for the influence of the relative humidity $f(h)$ and the temperature $g(T)$ (see Chapter 2.2.3). The calibration of the model consists on finding these two functions. In the following subsections the functions calibrated according to the experimental measurements are presented. The measurements conducted at 20 °C (see Chapter 4.1.1) served to calibrate the function $f(h)$ while the function $g(T)$ was calibrated by comparing these measurements with those conducted at elevated temperatures (see Chapter 4.1.2). Finally the function to calculate the hygrothermic coefficient of concrete K was formulated using the measurements conducted during the increments of temperature (see Chapter 4.1.3).

6.1.1 Drying at reference temperature

Assuming that the changes in the microstructure of the concrete due to drying can be neglected, the most important factor to determine the type of transport mechanism dominating the water transport process is the moisture content (see Chapter 2.2.1). Therefore, in a simplified diffusion approach to calculate water transport in concrete, the diffusion coefficient has to be define as dependent on the moisture content or rather on the relative humidity of the concrete pores. The approach from Bažant and Najjar [11] defines the influence of the relative humidity in the concrete pores on the diffusion coefficient by means of the relative humidity h (see Eq. 2.6 in Chapter 2.2.3). Following this approach, the parameters of the equation $f(h)$ can be calibrated to fit the measurements. With the parameters D_1 = 15 mm^2/d, α_0 = 0.02, n = 8, and h_c = 0.83 a good conformity between model and measurements can be obtained for the concrete MRC drying at 20 °C and 65 % RH (see continuous line in Fig. 6.1). However, if the concrete is subject to drying at a different environmental relative humidity, like for instance, 85 %, with the parameters calculated for 65 % RH the model does not follow the measurements accurately (see dashed line in Fig. 6.1).

According to the model from Bažant and Najjar, the diffusion coefficient is the higher, the higher the relative humidity of the concrete pores. As seen in Fig. 6.1, this approach implies that samples drying at higher environmental relative humidities reach the equilibrium with the environment faster than those drying at lower relative humidities. As it was discussed in Chapter 4.1.1.1 that could not be verified by the experiments. The experiments showed that the mean values of the relative humidity for concretes drying at 85 % and 65 % RH tend to reach the equilibrium with the environment at a similar point in time, as it would be the case if the diffusion coefficient were constant. The generality of the model as it was presented by Bažant and Najjar is very poor because it requires a new calibration for every time the relative humidity of the environment changes.

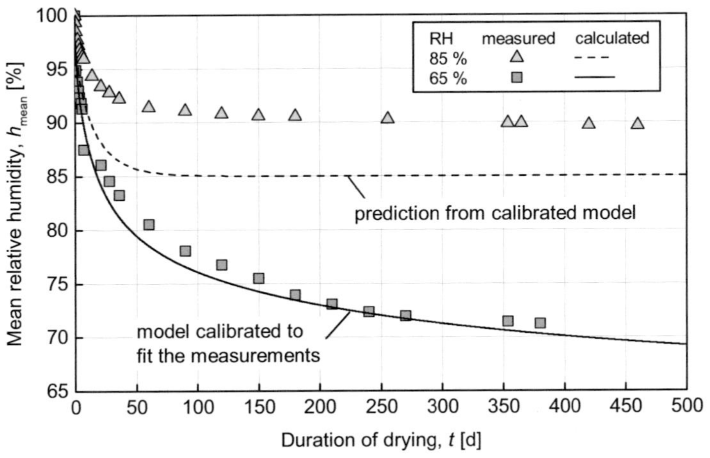

Figure 6.1: Comparisons between the measured values of concrete MRC drying at 20 °C and 85 % as well as 65 % RH and the calculated values using the model from Bažant and Najjar [11]

The lack of generality of the model can be overcome by including in the function that accounts for the influence of the moisture content on the diffusion coefficient of concrete two additional dependencies, namely the relative humidity of the environment and the gradient of the relative humidity. Hence, the function $\mathrm{f}(h)$ presented in Chapter 2.2.3 becomes $\mathrm{f}(h, h_\infty, dh/dr)$. After trying different configurations, the following definition was selected because of its simplicity and the good accuracy that the model provides when compared to the experimental measurements.

$$\mathrm{f}(h, h_\infty, \frac{dh}{dr}) = \frac{1}{1 + \left(\frac{1-h}{1-h_c(h_\infty, dh/dr)} \right)^n} \tag{6.3}$$

$$h_c(h_\infty, dh/dr) = 0.5 + \frac{0.5}{1 + 2 \cdot (1-h_\infty)^2} - 0.5 \cdot \left(\left| \frac{dh}{dr} \right| \cdot r_0 \right)^{0.5} \tag{6.4}$$

In the formulation presented above h_c is not a constant anymore but rather a function of the relative humidity of the environment h_∞ and the absolute gradient of the relative humidity $|dh/dr|$. This implies that the relative humidity at which the function drops down, ruled by the value of h_c, is dependent on the combined effects of the environmental relative humidity and the gradient of the relative humidity and moves towards lower values as the relative humidity in the concrete decreases. Furthermore, being the parameter h_c variable, the parameter α_0, as presented in Eq. 2.7 from Chapter 2.2.3, is not required and can be taken as zero which implies that the function $\mathrm{f}(h, h_\infty, dh/dr)$ decreases continuously as the relative humidity of the pores approaches 0 %. Eq. 6.4 includes the constant r_0 which equals 1 mm. In case the model is run in different dimensions than mm, this constant has to be converted to the correspondent units of length.

In case of drying at 20 °C, the model has two material dependent parameters, namely n (see Eq. 6.3) and D_1 (see Eq. 2.5 in Chapter 2.2.3). This model was compared with the measurements of relative humidity in the pore cavities conducted at 20 °C for all three concrete mixtures drying at 85 % and 65 % RH. Based on these comparisons the parameters n and D_1 were calibrated. Table 6.1 presents the calibrated parameters according to the w/c-ratio of the concrete samples.

Table 6.1: Parameters of the model calibrated according to the experimental results

Concrete Mixture	*w/c-ratio* [-]	D_1 [mm²/d]	n [-]
MLC	0.40	5.4	6.4
MRC	0.50	20.0	5.7
MHC	0.60	22.0	4.9

The generality of the model can be tested if after calibrating its parameters with the measured data, they can be approximated through defined functions. These functions have to be based on the characteristics of the concrete microstructure best described by the w/c-ratio. Eqs. 6.5 and 6.6 present the dependency of the parameters D_1 and n on the w/c-ratio of the concrete mix based on the calibrations conducted with the experimental results from the concretes MLC (w/c = 0.4), MRC (w/c = 0.5) and MHC (w/c = 0.6).

$$D_1 = 2.2 + \frac{19.8}{1 + (1.83 \cdot (1 - w/c))^{25}} \tag{6.5}$$

$$n = 6.1 - \frac{1.8}{1 + (2.3 \cdot (1 - w/c))^{9}} \tag{6.6}$$

The values of the diffusion coefficient at reference conditions D_1 are given in mm^2/d. According to Eq. 6.5, for concretes without capillary porosity (i.e. w/c-ratio < 0.4), D_1 approaches a constant value around 2.2 mm^2/d, while by concretes containing capillary porosity, D_1 increases steeply towards reaching a maximum value of 22 mm^2/d. The exponent n controls how steep the diffusion coefficient decreases with decreasing pore relative humidity and is the higher the lower the w/c-ratio of the concrete. In Eq. 6.6, n is described by a s-shaped function with a maximum value of 6.1, for w/c-ratios lower than 0.4, and a minimum value of around 4.3, for w/c-ratios higher than 0.7. The equations that compose the model of moisture transport are summarized in Appendix E.1.

In Fig. 6.2 the mean values of pore relative humidity calculated from the measurements conducted on the concretes MLC and MHC are compared with the calculations from the model using the parameters from Table 6.1.

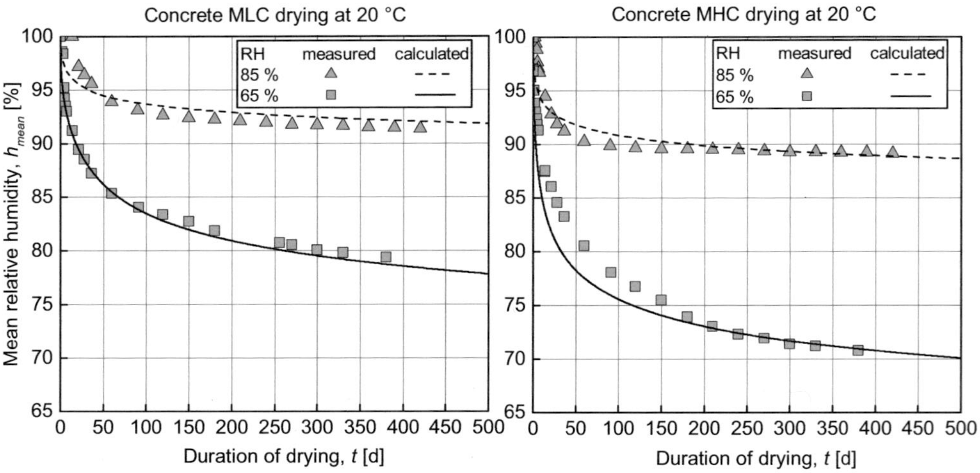

Figure 6.2: Comparisons between the calculated values using the model of moisture transport and the measured values of relative humidity from the concretes MLC (left) and MHC (right) drying at 20 °C and 85 % as well as 65 % RH

A very good concordance between model calculations and measurements is achieved for both concretes drying at 85 % and 65 % RH. Similar results were seen by the concrete MRC presented in Appendix D.1.

6.1.1.1 Comparison of the model calculation with experimental results from the literature

Very few measurements of pore relative humidity have been conducted and published in the literature, therefore a data base to further improve the calibration of the model and validate their results is not available. One of the few well documented experiments was conducted by Hanson [73] in the late sixties of the last century. Hanson recorded the development of the pore relative humidity of a concrete cylinder with a diameter of 150 mm drying at 20 °C and 50 % RH during three years. The w/c-ratio of the concrete mixture investigated by Hanson was 0.657. He used four Monfore relative humidity probes [99] and placed them in different positions over the radius to assess the humidity profile of the sample. The comparison of the model with the results from Hanson is presented in Fig. 6.3. The model can reproduce the behaviour of the concrete sample measured by Hanson very well even providing that the environmental relative humidity, w/c-ratio, and member size go beyond the ranges considered in the present investigation.

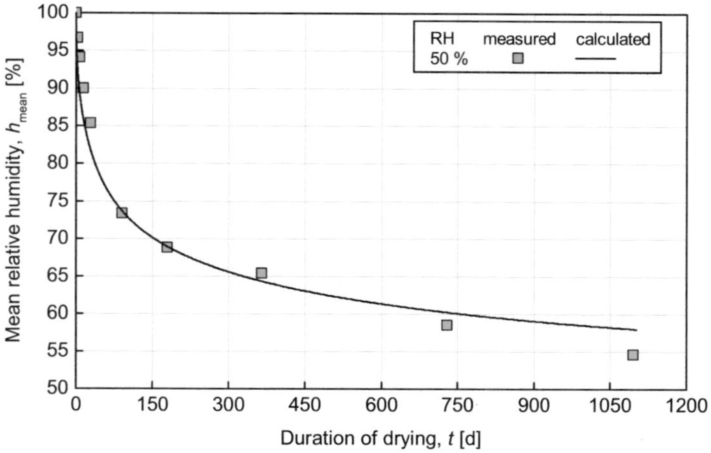

Figure 6.3: Comparisons between the calculated values using the model of water transport and the measured values of mean relative humidity from a concrete drying at 20 °C and 50 % RH according to Hanson [73]

Nevertheless, assuming that the function that accounts for the influence of the moisture content on the diffusion coefficient of concrete $f(h, h_\infty, dh/dr)$ is dependent on the relative humidity of the environment h_∞, implies that the diffusion coefficient which is an intrinsic property of the material is influenced by a condition that takes place outside the concrete microstructure. This is on first sight not quite correct, however, the environmental relative humidity denotes the final value of humidity to be reached in the concrete pores and under this consideration it is included in the calculation of the diffusion coefficient. Another aspect that needs to be discussed is the size of the concrete member. The dependency of the diffusion coefficient on the relative humidity of the environment as defined by Eq. 6.3 may vary considerably if the geometry of the concrete members to evaluate differs from the one used for the calibration. The diameter of the sample used by Hanson doubles the diameter of the samples used in the present investigation. However, even when the model shows very similar results when compared with the measurements from Hanson, a conclusive statement about the validity of the model for massive concrete members cannot be made.

6.1.1.2 Modelling concrete wetting

Although the measurements of concrete internal relative humidity were conducted only for drying specimens, the wetting of the concrete can also be calculated with the model. Under the assumption that the process of drying and wetting follow the same time dependency, Eqs. 6.3 and 6.4 must be changed as follows.

$$\mathrm{f}(h,h_\infty,h_0,\frac{dh}{dr}) = \frac{1}{1+\left(\frac{1-h_\infty+h-h_0}{1-h_c(h_0,dh/dr)}\right)^n} \tag{6.7}$$

$$h_c(h_0,dh/dr) = 0.5+\frac{0.5}{1+2\cdot(1-h_0)^2} - 0.5\cdot\left(\left|\frac{dh}{dr}\right|\cdot r_0\right)^{0.5} \tag{6.8}$$

Eq. 6.7 corresponds to the mirror image, with respect to a horizontal line cutting the function in two parts, of Eq. 6.3 which was formulated for drying. In order to model the process of wetting, the relative humidity of the environment h_∞ must be replaced by the initial relative humidity of the concrete pores h_0 for the definition of $h_c(h_0,dh/dr)$. Additionally, in the definition of $\mathrm{f}(h,h_\infty,h_0,dh/dr)$ the initial and end value of the relative humidity, h_0 and h_∞, need to be considered.

6.1.2 Drying at elevated temperatures

The drying process of concrete at elevated temperatures is accelerated in comparison to reference temperature. A well known mathematical expression to calculate the dependecy of the diffusion coefficient on the absolute temperature is the Arrhenius equation. Based on this equation, Bažant and Najjar [11] proposed the factor $\mathrm{g}(T)$ to calculate the influence of the temperature on the diffusion coefficient. Being 20 °C the temperature of reference, Eq. 2.7 from Chapter 2.2.3 becomes:

$$\mathrm{g}(T) = \frac{T+273}{293}\cdot\exp\left[\left(\frac{Q}{R_D}\right)\left(\frac{1}{293}-\frac{1}{T+273}\right)\right] \tag{6.9}$$

Assuming that the diffusion coefficient at elevated temperatures increases by a factor that obeys the Arrhenius equation, the quotient Q/R_D can be calculated for each of the concretes tested. After calibrating the diffusion coefficient of reference D_1 to fit the measurements conducted at elevated temperatures, they can be plotted in a $\ln(D_1)$ vs. $1/T_{\mathrm{abs}}$ diagram. The slope of the diagram corresponds to $-Q/R_D$. Following this procedure, the quotients Q/R_D were found for all three concrete mixtures and are presented in Table 6.2. According to the values from Table 6.2 the activation energy of diffusion Q varies with the concrete microstructure. At lower porosities, the activation energy is higher and decreases with increasing porosity of the concrete microstructure.

Table 6.2: Quotients between the activation energy of diffusion and the universal gas constant calibrated according to the experimental results

Concrete Mixture	w/c-ratio [-]	Q/R_D [K]
MLC	0.40	8000
MRC	0.50	4500
MHC	0.60	2500

Based on the three values from Table 6.2 calculated from the concrete mixtures, a function relating the w/c-ratio of the concrete mixture with the quotient Q/R_D was formulated. Eq. 6.10 describes a s-shaped relation where at lower w/c-ratios the quotient Q/R_D tends to a value of 9000 K and for w/c-ratios above 0.8, Q/R_D approaches 1800 K.

$$\frac{Q}{R_D} = 9000 \cdot \left[0.2 + \frac{0.8}{1+(2.1 \cdot w/c)^{10}}\right] \quad (6.10)$$

Fig. 6.4 shows a comparison between the results of the calculation using the model and the mean value of the relative humidity in the concrete pores measured for the concretes MLC and MHC drying at 60 °C and environmental relative humidities of 75 and 65 %.

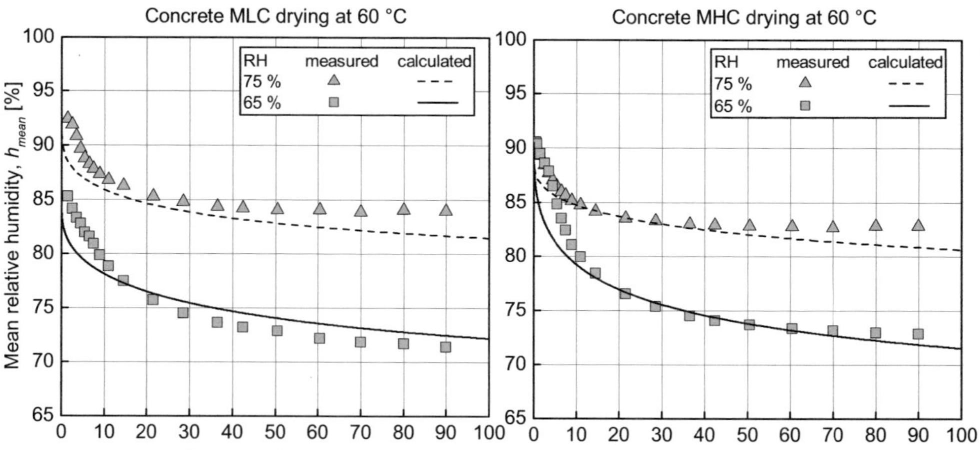

Figure 6.4: Comparisons between the calculated values using the model of water transport and the measured values of mean relative humidity from the concretes MLC (left) and MHC (right) drying at 60 °C and 75 % as well as 65 % RH

The model follows the development of the mean relative humidity in the concrete pores as it was measured for both concretes very accurately. Similar results were also achieved by the comparisons made for the concrete MRC drying at 60 and 80 °C (see Appendix D.1). As it was discussed in Chapter 4.1.2.2, at elevated temperatures, the differences between the drying processes of the concretes with low water content MLC and high water content MHC were less meaningful than the differences seen on the measurements conducted at reference temperature. As seen in Fig. 6.4, by the given conditions of elevated temperature and relative humidity, both concretes reached similar values of mean pore relative humidity within the duration of the measurements. According to the calculations conducted with the model, the reason for this behaviour can be found on the activation energy of diffusion. At elevated temperature, the concretes MLC and MHC behave similarly because the concrete MLC obtains more energy for the diffusion process than the concrete MHC due to the fact that the activation energy of the concrete MLC is higher. Even when the porosities of the concretes are different, at elevated temperature, the available energy allows the concrete MLC to reach a diffusivity comparable to the one that the concrete MHC reaches.

6.1.3 Effect of temperature changes

With the formulations presented in the past subsections, the relative humidity of concrete can be calculated for variable conditions of relative humidity of the environment at different but steady conditions of temperature. In order to develop a model capable of integrally calculating the drying and wetting process of concrete, the effect of variable temperature conditions must be taken into account. As it was briefly explained in Chapter 2.1.2.2, at constant moisture content, a change in the concrete temperature causes a variation of the relative humidity of the concrete pores. When the temperature raises, the water molecules within the concrete microstructure gain energy and the internal surfaces of the concrete microstructure cannot hold them anymore, causing a migration of the water molecules from small pores into larger pores. The consequences of this behaviour are described by the hygrothermic coefficient of concrete, which denotes the increment of the relative humidity in the concrete pores due to an increment in the temperature at constant moisture content.

Using the measurements presented in Chapter 4.1.3, a formulation for the hygrothermic coefficient of concrete K was developed. Eq. 6.11 describes a dome-shaped curve tilted to the right towards high relative humidities. The equation is calibrated with the parameters a, b and c, dependent on the concrete microstructure (see Eqs. 6.12 to 6.14).

$$K(h) = \frac{dh}{dT} = a \cdot (1-h)^b \cdot \exp\left[c \cdot (1-h)\right] \tag{6.11}$$

Eq. 6.11 must satisfy three important conditions. In first place, it has to be always positive because the relative humidity in the concrete pores always increases with increasing temperature as depicted by the comparison of the sorption isotherms of porous materials at different temperatures (see Chapter 2.1.2.2). The second and third conditions are given by the extreme values of relative humidity. If the vapour pressure of the concrete pores already equals the saturation pressure, any additional water coming from the smaller pores will condensate without elevating the relative humidity because it cannot be higher than 100 %. On the other hand, at relative humidities close to 0 %, no moisture is available to migrate from the small pores to the large pores and the value of K must be very close to zero. Eqs. 6.12 to 6.14 were derived to take into account the influence of the w/c-ratio on the hygrothermic coefficient of concrete.

$$a = 400 \cdot (w/c)^{14.3} + 0.15 \tag{6.12}$$

$$b = 12 \cdot (w/c)^{6.5} + 1.15 \tag{6.13}$$

$$c = -900 \cdot (w/c)^{11} - 7.0 \tag{6.14}$$

Eq. 6.11 in combination with Eqs. 6.12 to 6.14 allows to calculate the hygrothermic coefficient of concrete satisfying the conditions mentioned before and delivering plausible results for a range of w/c-ratios between 0.2 and 0.7. The comparisons of the calculated hygrothermic coefficient of concrete with the measurements conducted on sealed samples from the concretes MLC, MRC and MHC are presented in Fig. 6.5.

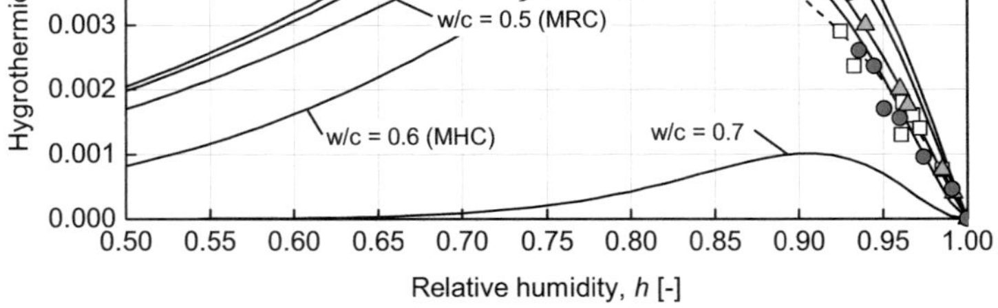

Figure 6.5: Changes of the relative humidity in the concrete pores due to a temperature change. Comparisons between calculated values and measurements

In Fig. 6.5 the calculations according to Eq. 6.11 are represented by the continuous lines, for which the corresponding w/c-ratios are indicated in the diagram. The dashed line represents the values of K calculated according to the formulation from Bažant and Najjar [11]. Following the tendencies shown by the measurements, the influence of a temperature change on the relative humidity of the concrete pores gain in importance as the relative humidity decreases from 100 % and reaches a maximum value by relative humidities around 85 %. The formulation from Bažant and Najjar assumes that this peak occurs at relative humidities around 70 %. Both formulations deliver values of the same order of magnitude, the difference lies on the inclination to the right that the present formulation assumes in order to fit the measured data in a better way.

The influence of the microstructure on the hygrothermic coefficient of concrete can be understood by means of the following considerations. As the gel pores are assumed to be completely filled with water while the capillary pores also contain air and water vapour, the relative humidity of the concrete corresponds to the relative humidity of the capillary pores. When temperature increases, the water molecules stored in the gel pores migrate to the capillary pores increasing their relative humidity. At low w/c-ratios, the amount of gel pores increases and the amount of capillary pores is reduced. The larger the amount of gel pores, the larger the amount of water molecules willing to migrate from the gel pores into the capillary pores, and moreover, the smaller the amount of capillary pores capable of storing these new water molecules. Therefore the hygrothermic coefficient of concrete has to be higher at lower w/c-ratios. This was verified by the measurements and taken into account in the formulation of the parameters from Eq. 6.11. In Fig. 6.5 besides the curves representing the w/c-ratios used for the concrete mixtures, two extreme values of the equation (w/c = 0.2 and w/c = 0.7) are plotted as well. At w/c-ratios lower than 0.4, the presence of capillary pores in the concrete microstructure can be neglected. Hence, the formulation for K assumes similar behaviours of the concretes with w/c-ratio lower or equal than 0.4. For concretes with w/c-ratio larger than 0.6, the amount of capillary pores increases in such a way that any increment in the relative humidity of these pores would need a considerable large amount of water molecules. As temperature raises, the water molecules coming from the gel pores will not be able to increase the relative humidity of the capillary pores in the same manner as concretes with lower w/c-ratios, because the amount of gel pores at such high w/c-ratios is restricted.

The presented formulation of the hygrothermic coefficient of concrete does not include any time influence and therefore it assumes that the changes in the relative humidity of the concrete pores occurs instantaneously after the variation of temperature takes place. As it was mentioned before, these changes obey a moisture exchange between gel and capillary pores which requires time to develop. In case of a temperature increase the water molecules gain energy which allow those located in the gel pores to abandon the surface at which they were bonded and head for zones of lower energy that can be found in the capillary pores. This moisture exchange between water molecules from the gel into the capillary pores takes place within a couple hours and therefore assuming an instantaneous change in the relative humidity of the concrete pores may be plausible with respect to the very much slower diffusion process of water molecules within the microstructure. Nevertheless, the opposite process, i.e. the movement of water molecules from the capillary pores into the gel pores due to a decrement in the temperature, occurs at a much lower energy state. The water molecules are attracted back to the surfaces of the gel pores through adsorption leading to a slow rearrangement of the water molecules located in the gel pores and therefore this process can take over weeks to reach equilibrium. In conclusion, the formulation of the hygrothermic coefficient of concrete as it is presented in Eq. 6.11 may be applied for increments of temperature even in cases where accounting for the time-development of the relative humidity is important. However, if a temperature reduction is considered the formulation can only indicate the final value of the relative humidity without providing any information about the time required to reach it.

6.2 Modelling the concrete mechanical properties

The formulations proposed in Chapter 5 allow to predict the mechanical properties of concrete members subject to elevated temperatures and different moisture contents. These formulations were based and calibrated according to the values of relative humidity measured in the concrete pores, which implies that their use is subjected to the condition of knowing the concrete relative humidity. On this background, in order to model the mechanical behaviour of concrete at varying temperature and environmental relative humidity, it is necessary to integrate the formulations of compressive and tensile strength, modulus of elasticity, shrinkage and creep from Chapter 5 with the water transport model from the present chapter in a new overall material model. This integration, i.e. development of the material model, is presented in the following subsections.

6.2.1 Assumptions

The objective of the new material model is to estimate the mechanical properties of concrete subject to changes of temperature and environmental humidity. The development of the material model was based in the following assumptions:

- The model is conceived for fully hydrated concretes. In case partly hydrated concretes are considered, any influence of further hydration on the mechanical properties is neglected.
- The concrete strength and stiffness of reference are defined at a relative humidity of 100 % (sealed conditions) and an ambient temperature of 20 °C.

- The influences of thermal incompatibility and drying on the strength and stiffness of concrete can be calculated separately and added to obtain the overall behaviour of the material.
- Thermal incompatibility produces irreversible damage to the concrete microstructure and its impact is strongly related to the water content of the concrete.
- Drying enhances the strength of concrete. This enhancement can however be reversed through wetting.
- The influences of hydrothermal reactions on the mechanical properties of concrete are not considered.
- The processes of drying, drying shrinkage and drying creep of concrete obey the same time-development function.
- Basic creep is dependent on the moisture content of the concrete, which implies that changes in the moisture content also influence the development of basic creep.

6.2.2 Implementation

Concrete has a thermal diffusivity of around 86000 m^2/d while the moisture diffusivity, according to the water transport model, lies between 20 and 700 m^2/d depending on the temperature. Due to the enormous difference between the diffusivities and the fact that the concrete samples had a diameter of only 75 mm, solving the heat transport equation may not be necessary. For the calculations, the temperature is rather given as input in a time dependent function assuming a constant temperature over the geometry of the concrete member. For the implementation of the material model, a computer program designed to run in the software MATLAB was developed. The program integrates the solution of the diffusion equation (see Eq. 6.1), the equation of the hygrothermic coefficient of concrete (see Eq. 6.11) and the formulations from Chapter 5. The program consists of a main routine called *Tempecon* and four subroutines. Subroutine *Middle* which calculates the mean value of a given parameter over the geometry of the concrete member, subroutine *Rand* used to check the conditions of temperature, relative humidity of the environment and load in every time step, subroutine *Refhum* which calculates the equivalent relative humidity at reference temperature by solving the equation of the hygrothermic coefficient of concrete and subroutine *Watdiff* used to calculate the moisture diffusion in the concrete member. The flow charts and scripts of the subroutines are given in Appendix F.2.

6.2.3 Coverage and capabilities

The coverage of the new material model is limited according to the formulations on which the model is based. In general, the formulations to predict the mechanical behaviour of concrete presented in Chapter 5 as well as the model of water transport depend on one mixture property and two state variables, namely w/c-ratio and temperature and related relative humidity of the concrete pores. The formulations were conceived based on the laboratory tests conducted on the concrete samples and therefore it is expected that the accuracy of the predictions from the material model is higher for concretes with similar mixtures and geometries subject to similar conditions of temperature and environmental relative humidity as those used in the laboratory investigations. However, the formulations are supported on physical sound assumptions and it may be possible to use the model to calculate the behaviour of concrete members with geometries, w/c-ratios, conditions of relative humidity and temperature above and below the ones tested. The material model can be used to predict the behaviour of the tensile and compressive strength, modulus of elasticity, drying shrinkage and creep for concrete mixtures and environmental conditions within the ranges presented in Table 6.3.

Table 6.3: Coverage of the new material model

Sample properties and environmental conditions		Ranges of coverage	
		general	investigated
w/c-ratio	[-]	0.3 - 0.8	0.4 - 0.6
Member size	[mm]	arbitrary	75
Relative Humidity	[%]	30 - 100	40 - 100
Temperature	[°C]	20 - 100	20 - 80

In Table 6.3 the general range of coverage of the model is indicated on the left column. The column on the right indicates the ranges investigated in the laboratory tests for which the accuracy of the model is expected to be higher.

Besides defining the ranges of concrete mixtures and environmental conditions that the material model can cover, other important aspects to consider are the capabilities of the model. The material model reproduces the behaviour of the concrete mechanical properties based on the relations between temperature and moisture content that were formulated in Chapter 5. The time-development of the temperature is given as input while the time-development of the moisture content is calculated by the model. Therefore, the capabilities of the material model are limited by the capabilities of the model of water transport. The model of water transport presented in the beginning of the present chapter is capable of calculating the moisture movement in concrete for drying and wetting including the action of temperature increments. Due to the limitations on the definition of the hygrothermic coefficient of concrete (see Section 6.1.3), the time-development of the relative humidity of the concrete due to a change of the temperature cannot be assessed by the model, i.e. the movement of water molecules from gel to capillary pores and vice versa. This fact has major consequences on the capabilities of the model that may vary on importance according to the mechanical property that wants to be predicted.

In case of an increase of the temperature, the water molecules move from the gel into the capillary pores within a couple hours. Not knowing the time development of this process barely affects the results of the model regarding concrete strength and stiffness, however, in case of creep, it leads to an underestimation of the creep deformations that occur during the first hours of loading. This statement can be understood with help of the following considerations. The model assumes that the increment of temperature and the associated increase of the relative humidity of the capillary pores occur simultaneously, neglecting the delay taking place between these two processes. Therefore, assuming that drying creep may develop due to the movement of water molecules in the concrete microstructure while loading, the creep deformations produced by the movement of water from gel pores into capillary pores that occur after reaching a higher temperature is not considered by the model because the model assumes that this water movement is achieved before the loading begins.

In case of reducing the temperature, the water transport between capillary pores and gel pores can take over weeks. Without including the time development of this process in the equation of the hygrothermic coefficient of water it is not possible to calculate any further development of the mechanical properties. By assuming that thermal incompatibility produces irreversible damage to the concrete microstructure, the model can only give an estimation of the final values of strength and stiffness that will be reached after a reduction of the temperature have taken place. On the other hand, no fair results for creep and shrinkage may be expected because the model would decrease the relative humidity immediately, affecting the calculation of both the basic and drying components of concrete creep.

6.2.4 Comparing the model results with measurements

The main experimental results from Chapter 4 were used to conceived the formulations that compose the material model and some other obtained results can as well be used to test its accuracy, i.e. to validate it. Some comparisons between test results and calculations with the material model are presented below. The equations that constitute the developed model are summarized in Appendix E.

6.2.4.1 Strength and Stiffness

The tests of strength and stiffness were conducted on the concrete samples following defined conditioning schemata (see Chapter 3.2.1). In Fig. 6.6 the model is tested according to the conditioning scenario at which some samples of the concrete MRC were subject. Starting with a reference temperature of 20 °C and a mean relative humidity of 100 %, the samples were subject to drying during 100 days at 20 °C and 65 % RH. Then they were heated up to 80 °C and kept at a relative humidity of 65 % for 90 days longer. The conditioning process is indicated on the top figure of the diagram.

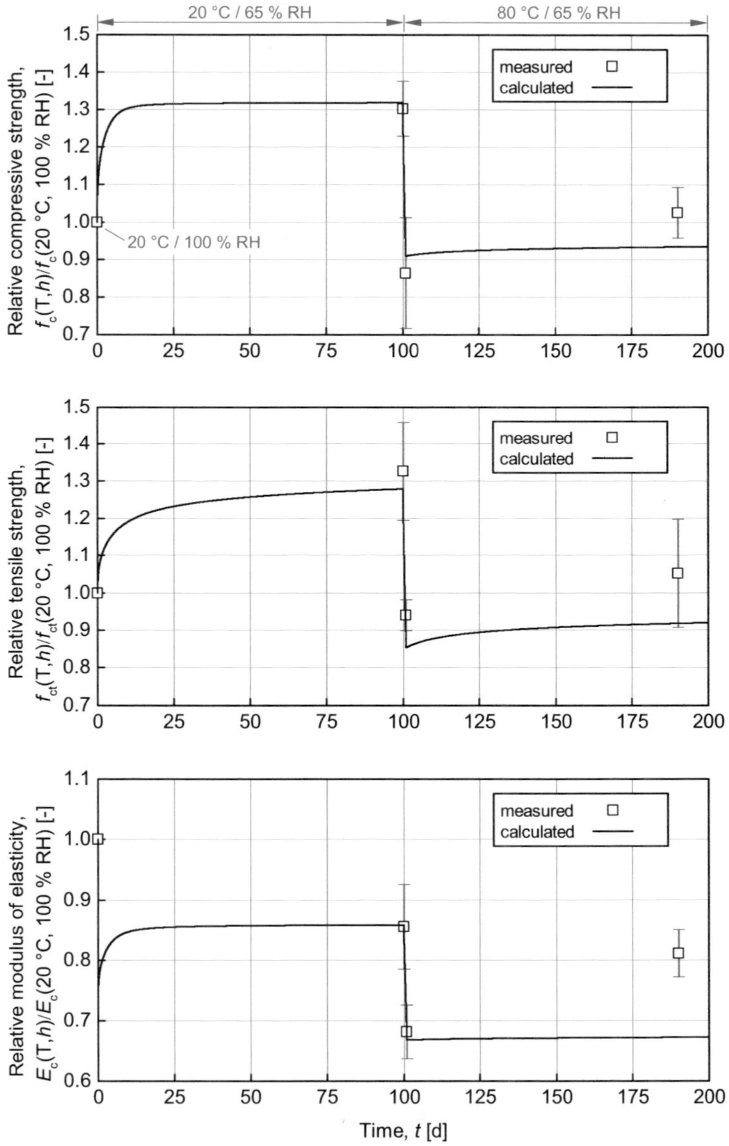

Figure 6.6: Comparison between calculations with the new material model and measurements of compressive strength (top), tensile strength (middle) and modulus of elasticity (bottom) for the concrete MRC after drying during 100 d at 20 °C / 65 % RH and then heated during 100 d at 80 °C / 65% RH

Measurements of the mechanical properties were carried out at reference conditions, after the end of the drying period at 20 °C, shortly after increasing the temperature and at the end of the period of sustained elevated temperature. In Fig. 6.6 the compressive strength, tensile strength and modulus of elasticity of the concrete relative to the properties at reference conditions are plotted with time according to the conditioning process. The measured values are depicted by the square icons including an indication of the scattering of the experimental results by displaying, between horizontal bars, the standard deviation. The results of the material model represented by the continuous lines follows quite well the experimental results. It reproduces the increment on the compressive and tensile strength due to drying during the first 100 days, the sudden change in these strengths due to the increment of temperature from 20 to 80 °C and the increase of the strengths due to further drying afterwards. Unlike the strength development, in case of the modulus of elasticity, the model calculates an abrupt change in the concrete stiffness at the beginning of drying followed by an increment of it as the drying continues. In Appendix E.2 the equations of the material model regarding the influence of temperature and moisture content on the strength and stiffness of concrete are summarized.

For the conditioning scenario considered in Fig. 6.6, the model is able to reproduce the behaviour of the concrete very accurately. However, other condition scenarios involving higher moisture content during the heating process cannot be reproduced by the model in the same way. An example of it is presented in Fig. 6.7. This figure presents the comparison between model calculations and measurements of tensile strength and modulus of elasticity from samples that were stored at 20 °C and 95 % RH during 200 days and then heated up to 60 °C while the relative humidity was kept at 95 %.

At high ambient humidities the concrete loses only a small amount of moisture to the environment and therefore the enhancement of the strength as well as the recovery of the modulus of elasticity after dropping down during drying are limited. In Fig. 6.7 it can be seen how the model reproduces the behaviour of the concrete during the drying process at 20 °C very well. Nevertheless, once the temperature is increased, the measured results differ from the prediction of the model considerably. This is due to the fact that the model neglects the influence of hydrothermal reactions. In case of compressive or tensile strength the hydrothermal reactions cause a healing of the microcracks produced by the thermal incompatibilities and therefore, unlike the model, the measured values do not show a decrease of the strength after heating. The healing of microcracks has however smaller effect on the concrete stiffness and in consequence, the model behaves better when predicting the behaviour of the modulus of elasticity.

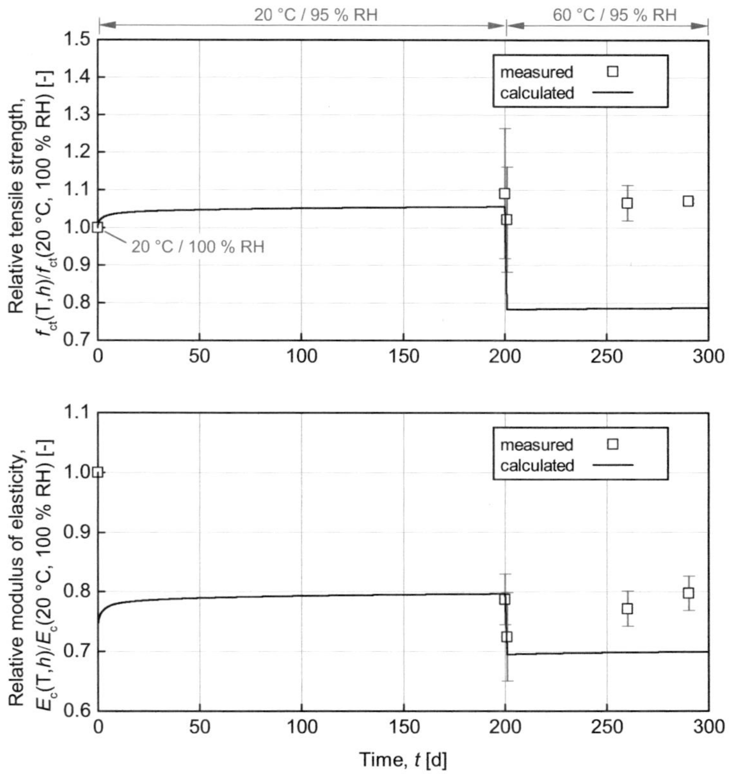

Figure 6.7: Comparison between calculations with the material model and measurements of tensile strength (top) and modulus of elasticity (bottom) for the concrete MRC after drying during 200 d at 20 °C / 95 % RH and then heated during 100 d at 60 °C / 95% RH

The comparisons presented in Figs. 6.6 and 6.7 correspond to two extremes in terms of the accuracy of the model. At lower moisture content the model delivers very accurate predictions because the influence of the hydrothermal reactions is negligible which is in accordance with the assumptions of the model. Meanwhile, by concretes containing a high amount of moisture, the hydrothermal reactions play an important role on the strength development that the model cannot reproduce. Further comparisons including other conditioning scenarios at which samples from the concrete MRC were subject are included in Appendix D.

6.2.4.2 Time dependent deformations

In the following diagrams results from the material model are compared with some selected measurements of shrinkage and creep. Additional comparisons including all the creep and shrinkage measurements conducted on samples of the three concrete mixtures are presented in Appendix D.3 and D.4. In order to predict the shrinkage and creep deformations of the samples, the model has to recreate the whole history of storage conditions to which the samples were subject before being loaded. Previously to conducting the creep tests, the samples used for the measurements were conditioned for over 450 days at 20 °C and relative humidities of 95, 85 and 65 %. Moreover, the samples were loaded after the testing temperature in the chamber was reached (see Chapter 3.2.5).

Fig. 6.8 compares the shrinkage deformations measured on the concrete MHC during the creep tests conducted at 20 and 70 °C and 65 % RH with the calculations of the material model. The equations of the material model regarding shrinkage and creep are summarized in Appendix E.3 and E.4 respectively.

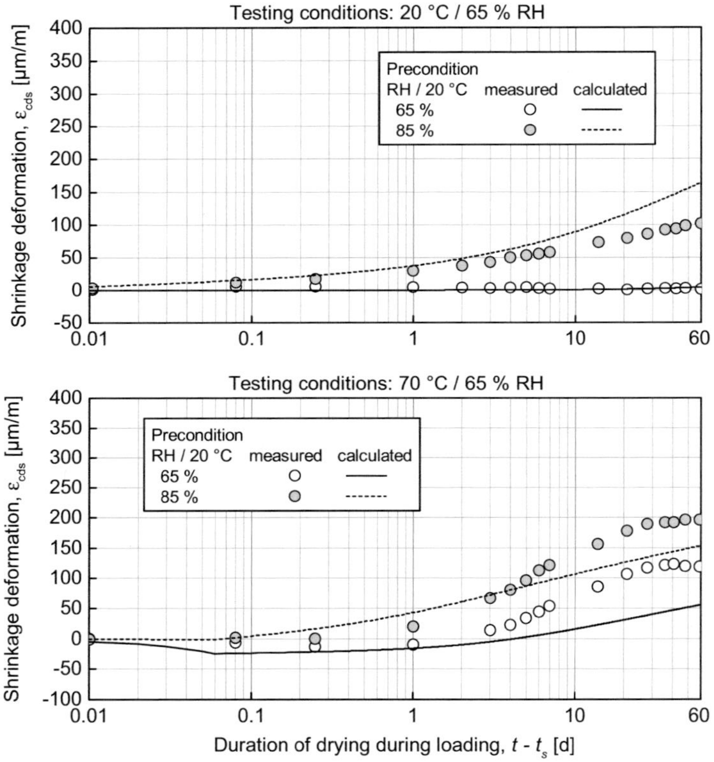

Figure 6.8: Drying shrinkage calculated with the material model in comparison with the measured values for the concrete MHC during the creep tests conducted at 20 °C (top) and 70 °C (bottom) and 65 % RH

In Fig. 6.8 the shrinkage deformations are plotted against time. The time axis indicates the duration of loading which means that even when the material model is able to calculate the shrinkage deformations from the beginning of the drying process, only the shrinkage deformations occurring after applying the load were taken into account for the comparisons with the measurements.

As it can be appraised on the top part of Fig. 6.8, the material model is able to predict the shrinkage behaviour of the concrete during the creep tests at 20 °C very well. For the samples that were previously conditioned at 65 % RH, the model predicts barely no change in the shrinkage deformations, because after 450 days of drying at 65 % RH, only small changes on the mean relative humidity may still occur if the ambient relative humidity remains unchanged. This was found to be in perfect accordance with the measurements. The samples previously conditioned at 85 % RH can very well lose moisture to an environment with 65 % RH, and therefore the material model predicts the development of shrinkage deformations following the tendency shown by the measurements. The bottom part of Fig. 6.8 presents the comparisons between the calculations with the material model and the measurements of shrinkage during the creep tests conducted at 70 °C. The development of the shrinkage deformations as presented by the material model seems to be slower in comparison with the measurements. This may be due to the fact that the air blowers within the climate chambers accelerated the drying process during the creep tests (see Chapter 3.2.5). This acceleration is not taken into account by the model because the model of moisture transport was calibrated based on measurements conducted without any influence of air blowers.

The creep deformations measured in the concrete MRC during the tests conducted at 20 and 70 °C and 65 % RH are compared with the model calculations in Fig. 6.9. For the concrete MRC the samples were conditioned at three different ambient relative humidities before conducting the creep tests, namely 95, 85 and 65 %. The diagram on the top corresponds to the creep tests conducted at 20 °C and 65 % RH and the one on the bottom contains the results from the tests conducted at 70 °C and 65 % RH.

The model can reproduce the influence of moisture content on the development of creep. For the creep tests conducted at 20 °C, the model follows the tendency of the measurements. The higher the moisture content at the beginning of loading, the higher the creep deformations that develops. In case of the tests conducted at 70 °C, the model also recognises the importance of the initial water content on the development of the creep deformations. A rather surprising good prediction is shown by the model for the samples previously conditioned at 95 % and 85 % RH. Both model and measurements show almost the same creep deformations for these samples. Following the definition of the hygrothermic coefficient of concrete (see Section 6.1.3), after increasing the temperature from 20 to 70 °C, the relative humidities reached by the samples previously conditioned at 95 % and 85 % are barely the same, and therefore the drying and basic creep that develops afterwards may also be similar.

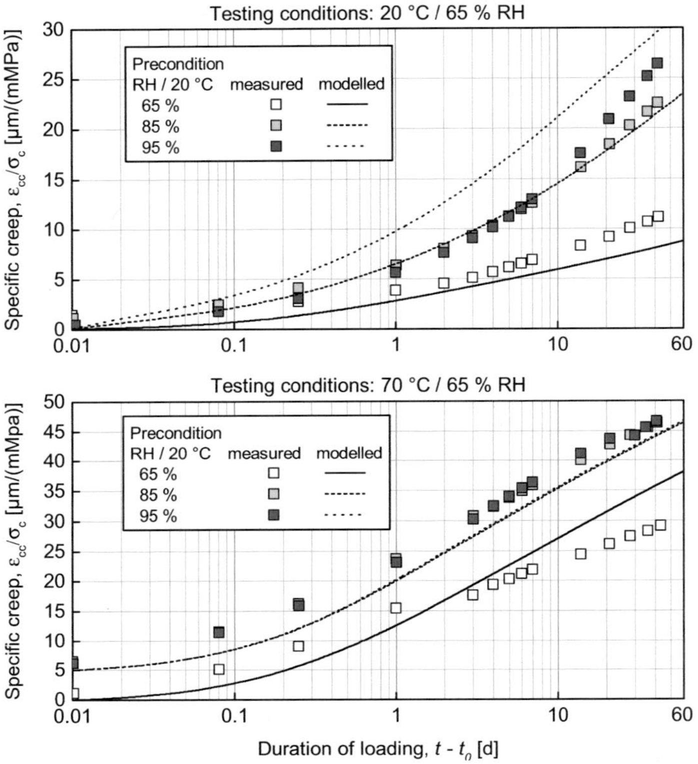

Figure 6.9: Specific creep calculated with the material model in comparison with the measured values for the concrete MRC during the creep tests conducted at 20 °C (top) and 70 °C (bottom) and 65 % RH

In Section 6.2.3 the capabilities of the model regarding the effect that increasing the temperature has on creep were discussed. It was mentioned that, in case the samples are loaded after being heated, the model underestimates the creep deformations that occur during the first hours of loading because in the formulation of the hygrothermic coefficient of concrete no time-development function is considered. In Fig. 6.9 as well as all figures presented in Appendix D.4, where elevated temperatures are involved, the curves were intentionally dragged up to match the initial values at $t - t_0 = 0.01$ d. In this way, it is shown how the material model follows the tendency of the creep deformations. It is known that, by neglecting the time dependency of the hygrothermic coefficient, the model is not able to reproduce the development of creep during the first hours after loading properly. Nevertheless, even though the results are affected by this weakness, the material model not only follows the tendency of the creep development but also achieves a quite good accuracy in comparison with the values of the tests results.

6.3 Summary and conclusions

In first place, a model capable of calculating the water transport in cylindrical concrete samples subject to drying at any condition of temperature and environmental relative humidity and even being able to include variable conditions over time was presented. Subsequently a material model to couple the water transport model with the formulations from Chapter 5 was implemented in a program based on the finite difference method. Finally, the validity of the calculations carried out with the program was tested by comparing them with the experimental results.

The process of development of the model of water transport contributed with the comprehension of the physical background of the problem. The most important conclusions derived from this development are subsequently summarized:

- The drying process in concrete is so complex that a simple dependency of the diffusion coefficient with the relative humidity is not enough to model it accurately. It was necessary to include, additionally, the relative humidity of the environment and the gradient of relative humidity in the concrete in order to adequately model the moisture transport in concrete.
- Under the assumption that drying and wetting of concrete follow the same time-development function, two definitions for the diffusion coefficient had to be proposed, one to model drying and another to model wetting.
- Diffusion at elevated temperatures is improved by the effect of the activation energy of diffusion. The activation energy is dependent on the concrete microstructure. For concrete mixtures with low w/c-ratios, the activation energy is higher than for concretes with higher w/c-ratios. This implied that, at a temperature of 60 °C, the concretes tested reached similar diffusivities independent of the w/c-ratio of the mixtures, even though low w/c-ratios are associated with a denser microstructure.
- The hygrothermic coefficient of concrete is dependent on the relative humidity of the concrete pores reaching maximum values in the range between 80 % and 90 % RH. It is also higher for concretes with lower capillary porosity.
- The simulation of moisture transport in concrete involving reduction of the temperature cannot be carried out with the presented model. It would be necessary to reconsider the formulation of the hygrothermic coefficient of concrete. The present formulation neglects the time dependency of the changes of the relative humidity in concrete due to variations in the temperature, which may be valid in case of increasing temperature but not if the temperature is reduced.

The implementation of the material model, i.e. the coupling of the newly developed moiture transport model with the basic equations developed in Chapter 5, and further comparisons of the calculations with the results of the conducted experiments related to the mechanical behaviour of concrete also led to some important conclusions.

- The material model can calculate the time behaviour of the strength and stiffness of concretes subject to elevated temperatures and different environmental relative humidities.
- The accuracy of the material model varies by the calculations of compressive and tensile strength. The accuracy of the model is higher by concretes that are heated while having low moisture content. By concretes with a high amount of moisture in their microstructure, the model neglects the influence of the hydrothermal reactions which leads to a overestimation of the final effect of temperature. The values of strength determined by the model in these cases are however on the safe side.
- The influence of the hydrothermal reactions on the concrete stiffness is secondary, and therefore the accuracy of the material model, for the prediction of the modulus of elasticity, is not affected by the moisture content of the concrete to be evaluated.
- Due to the process of conditioning to which the samples were subject before the creep tests were conducted, only a model capable of reproducing the entire history of conditioning, as been developed here, can accurately calculate the time dependent deformations measured on the samples.
- The material model is able to differentiate between the basic and drying components of the shrinkage and creep deformations. It assumes that the total shrinkage and creep deformations correspond to the summation of their basic and drying components. However, in case of shrinkage, the developed model can only calculate the effects of elevated temperatures and changes in the moisture content on the drying component of shrinkage. Basic shrinkage could also be included in the model by adopting the formulations from Model Code 2010 [N15]. Nevertheless, this was not accomplished here because no experiments regarding basic shrinkage were conducted.
- The material model reproduced the tendencies of the creep and shrinkage behaviour shown by the samples tested at 20 °C as well as at elevated temperatures. The samples previously conditioned at lower relative humidities showed lower shrinkage and creep deformations due to the influence of the moisture content at the time of loading.
- The material model was able to predict and explain the reasons for the similar behaviours of the creep deformations seen between the samples from the concrete MRC previously conditioned at 20 °C and 95 % and 85 % RH and tested at 70 °C and 65 % RH. According to the formulation of the hygrothermic coefficient of concrete

presented in the present chapter, the similarity between the creep deformations of these samples is due to the fact that both reached similar relative humidities after the temperature was increased.

- A weakness of the material model regarding the prediction of creep was identified by the comparisons with the tests conducted at elevated temperatures. This weakness resides in the definition of the hygrothermic coefficient of concrete. The material model is not able to adequately predict the behaviour of creep during the first hours after the temperature is increased because no time-development function was included in the definition of the hygrothermic coefficient of concrete.

Chapter 7

Summary and outlook

The material ageing of concrete is generally associated with changes over time in mechanical properties such as modulus of elasticity, compressive and tensile strengths and creep, but as well physical properties like e.g. shrinkage. This thesis describes the process of development of a new overall material model with the capabilities required to evaluate the concrete ageing due to the influences of the environment. The model developed is able to predict the time-dependent behaviour of the compressive strength, tensile strength, modulus of elasticity, creep and shrinkage of concrete subject to elevated temperatures up to 100 °C and variable ambient relative humidity.

The conception of the material model followed a rigorous process of evolution. The initial objective of this investigation was to develop a model to predict the mechanical behaviour of concrete subject to elevated temperatures up to 100 °C. Nevertheless, as the investigation progressed, the need to incorporate the moisture content in the approach of the model became evident. By reviewing the current knowledge about the influence of temperature on the mechanical properties of concrete, it was clear that the presence of water within the microstructure of the hardened cement paste is crucial to understand the behaviour of the concrete mechanical properties. In addition, it was found that elevated temperatures affect the mechanical properties of concrete not only originating damage to the concrete microstructure but also influencing the water storage capacity of concrete and promoting the exchange of moisture between the pores of the microstructure and the surroundings, which could create positive effects. For these reasons the original idea of the model had to be re-evaluated and modified. The model shall not only predict the influence of temperature but also the influence of moisture content.

The material model was proposed after a long process of development that can be divided in four phases. First, a literature review was made to settle the basis of the investigation. Then, experiments were designed and conducted to gather information and solve open questions. Central features of the experiments were the spatially resolved detection of temperature and humidity across the cross section of different samples from 3 different concretes and the development of experimental equipment to condition the samples and to measure concrete creep at elevated temperatures while maintaining a constant relative humidity.

The experimental results allowed the formulation of widely physically based mathematical expressions to relate the conditions of moisture content and temperature with changes on the mechanical properties. Finally the material model was developed. The following sections summarize the main findings achieved along the investigation and the open questions that arose in the process.

7.1 Main findings of the thesis

The new overall material model has three main components:

1. A model to calculate the moisture transport in concrete.
2. Mathematical formulations to calculate the influences of elevated temperature and pore relative humidity on the strength and stiffness of concrete.
3. Mathematical formulations to calculate creep and shrinkage at elevated temperatures and different ambient humidities.

The development of each of these components generated new knowledge and expertise that can be summarized in the following findings.

Findings related to the development of the model for moisture transport

- The measurements showed that the drying process of concrete is accelerated by the influence of temperature and is the slower, the lower the porosity of the concrete microstructure. However, at elevated temperatures, the differences seen in the water transport properties between concretes with low porosity and concretes with high porosity were less significant than those seen in the same concretes tested at 20 °C. This issue could be clarified by the development of the model of water transport. At elevated temperatures, the diffusivity of the concrete is improved by the effect of the activation energy of diffusion. According to the experiments and the assumptions made for the model, this energy is the higher the lower the porosity of the concrete microstructure, and therefore the concretes tested at elevated temperatures reached similar diffusivities independent of the porosities of their microstructures.
- The diffusion coefficient of concrete is neither constant nor dependent only on the moisture content of the concrete, as it is commonly assumed in the literature. The measurements of relative humidity in the concrete pores showed that concretes, independently of the relative humidity of the environment, tend to reach the equilibrium with the surroundings after similar time periods. This behaviour can be reproduced in a model based on diffusion by using a constant diffusion coefficient; but is not possible to reproduce it if a diffusion coefficient dependent on the

relative humidity is used. However, the measurements and further calculation with the water transport model showed that the time-development of the relative humidity of the concrete pores during drying is highly non-linear, which discards the possibility of using constant diffusion coefficients. In order to model the moisture transport in concrete accurately, it was necessary to include in the formulation of the diffusion coefficient, dependencies on the pore relative humidity, the ambient relative humidity and the gradient of the relative humidity within the concrete.

- As denoted by the hygrothermic coefficient of concrete, the measurements confirmed the increase of the relative humidity in the concrete pores due to an increment of the temperature. The formulation proposed for the hygrothermic coefficient of concrete was based on the conducted measurements and suggests that this coefficient is the lower, the higher the porosity of the concrete and is also dependent on the relative humidity of the concrete pores, reaching maximum values in the range between 80 % and 90 % RH.

Findings related to the development of the formulations for concrete strength and stiffness

- The experimental program was designed under the assumption that the influence of the environment on the mechanical properties of concrete can be explained by superposition of the effects that temperature and moisture content cause on the concrete mechanical properties. Assuming superposition implies that no additional effects resulting from the interaction between temperature and moisture are considered. The measurements suggested that this assumption do not apply to concretes with high moisture content. A third effect known as hydrothermal reactions emerges from combining high temperature and high moisture content in the microstructure of a concrete made with siliceous aggregates. By means of the conducted literature review, it was known in advance that hydrothermal reactions were feasible to occur. The consulted sources suggested that the influences of these reactions on the concrete strength and stiffness were only secondary. However, the experimental results showed that subjecting the concrete samples to elevated temperatures while having a high amount of moisture content in their microstructures influences the tensile and compressive strength considerably. Nevertheless, the effect of the hydrothermal reactions on the modulus of elasticity of the concrete samples was less significant.

- Both compressive and tensile strengths increase with decreasing water content in the concrete microstructure. By comparing the results from concretes of different porosities, it was found that the influence of drying on the concrete compressive and tensile strengths is not dependent on the porosity of the concretes.

- Drying has a negative influence on the modulus of elasticity of concrete. The experimental results showed that samples stored at 20 °C and 100 % RH had the highest modulus of elasticity. After a small change in the relative humidity of the concrete pores took place, an abrupt change in the stiffness of the samples was measured and as drying continued, the stiffness partially recovered.

- The experiments showed that after subjecting a concrete sample to elevated temperatures, an instantaneous and negative effect on the strength and stiffness of the material takes place. Any recovery of these mechanical properties with time must be accompanied of a loss of moisture or associated with further hydration or the development of hydrothermal reactions. These processes are however intensified by the effect of elevated temperatures.

- The developed material model is able to predict the combined effects of elevated temperatures and moisture content on the strength and stiffness of concrete in the ranges between 20 and 100 °C for environmental humidities above 30 % RH.

- Based on the experimental observations and taking into account the equations of the material model, the following conclusions can be drawn:

  The concrete compressive strength can increase up to 32 % for concretes with low moisture content (i.e. around 50 % RH) compared to moist concretes with pore relative humidities close to 100 % RH. In case of tensile strength this increment amounts to approximately 40 %.

  Elevated temperatures influence the concrete strength and stiffness negatively. For the range between 20 and 100 °C, in average the compressive strength and tensile strength decrease with increasing temperature at a rate of 5 % every 10 K while the modulus of elasticity decreases 3 % every 10 K.

Findings related to the development of the formulations for concrete creep and shrinkage

- The moisture content exerts a strong influence on the creep and shrinkage deformations. The experiments showed a clear dependency of the development of creep and shrinkage with the initial content of water of the concrete samples. The samples previously conditioned at lower relative humidities developed small shrinkage and creep deformations due to the lower moisture content present in their microstructure.

- The experimental results showed that, independently of the initial content of moisture and the temperature of the concrete samples, the rate of change of the mean relative humidity scales in a unique manner the rate of drying shrinkage. According to the developed formulations, drying shrinkage and mean relative humidity follow the same kinetics. Moreover, given the fact that basic creep is influenced by the moisture content of the concrete, the formulations suggest that drying creep also follows the same kinetics as the mean relative humidity.

- The results of the creep tests showed that the total creep deformation increases with increasing temperature. Furthermore, the calibrated formulations for creep of the new material model allowed the identification of different effects of the temperature on the development of basic and drying creep. According to these formulations, the component of basic creep increases with increasing temperature and the component of drying creep reaches, although faster, lower final values at elevated temperatures than at 20 °C.
- The developed material model is able to calculate the time dependent deformations of concrete taking into account the history of conditioning before the load application takes place for temperatures between 20 and 100 °C. For this purpose, it has been mandatory to include the differentiation between the basic and drying components of shrinkage and creep in the material model.
- By analysing the results of the conducted experiments and the outputs of the material model the following conclusions can be drawn:

  The basic creep deformations are in average twice as high at 30 °C than at 20 °C. Basic creep is three times higher at 70 °C compared to 20 °C.

  Although the final values of the drying creep and drying shrinkage deformations are strongly dependent on the initial moisture content of the concrete, they decrease with temperature in average at a rate of 5 % every 10 K.

7.2 Open questions

In order to reach the goal of this investigation, a significant but not unlimited amount of time and resources were employed. Therefore, some important issues were intentionally left outside the scope of the investigation. In addition, not all questions that emerged during the investigation could be completely clarified. All these aspects are finally reflected on the new overall material model that was developed by limiting its applicability and coverage. The clarification of the following facts is therefore reserved for future research work.

All concrete samples were made using siliceous aggregates. This means that the influence of other types of aggregates cannot be described by the model. For aggregates with considerably lower thermal expansion coefficients like, for instance, calcareous aggregates, the formulations that relate changes in the temperature with changes in the strength and stiffness of the concrete may not be valid. However, based on additional related experimental investigations, the new model may be easily adapted.

It has been assumed in the investigation that drying was the only process ruling the kinetics of the mechanical behaviour of the samples subject to elevated temperatures.

However, for samples subject to elevated temperatures and high humidities, hydrothermal reactions also play an important role. In the experiments, neither the influence of the hydrothermal reactions on the compressive and tensile strengths of concrete nor the way they develop on time were determined. This would have blown up the scope of the present work. For these reasons, the material model does not consider the effects of hydrothermal reactions. This implies that the model overestimates the influence of elevated temperatures when it considers concretes with a high amount of moisture in their microstructure. Nevertheless, the developed material model is suitable of being adapted to integrate the described effect, if this effect is quantified experimentally.

Based on the measurements, a formulation to calculate the influence of a temperature change on the relative humidity of the concrete pores was developed. The mathematical expression proposed for the so-called hygrothermic coefficient of concrete determines that the pore relative humidity increases with increasing temperature and decreases if the temperature is reduced. However, it does not include the time development of these processes, which implies that they are assumed to be instantaneous. This has consequences on the applicability of the material model. The model can neither calculate the time development of the concrete strength and stiffness nor the creep and shrinkage when a reduction of the temperature takes place because the associated process of reduction of the relative humidity in the concrete pores takes actually several days. However, the model can estimate the values that the mechanical properties will reach after the process of reduction of the relative humidity is completed, which is decisive in most practical cases. For the case when the temperature increases, the assumption of instantaneous increment of the relative humidity is less inaccurate. This process takes only a couple hours, which does not affect the calculation of the strength and stiffness but reduces the accuracy of the creep model as the associated effects on the creep deformation during the first hours after loading cannot be considered.

In the creep tests conducted at elevated temperatures the samples were loaded after reaching the testing temperature. Even though the material model is capable of calculating the influence of heating during loading on the creep deformations, such cases were not investigated and therefore no own-conducted experiments were available to verify the results of the model. However, a first orienting comparison with corresponding results from the literature showed that the model also reflects the effect of transient creep accurately.

The process of wetting of concrete was approached only theoretically. In order to verify the definition proposed for the diffusion coefficient in case of wetting, appropriate experimental measurements need to be conducted. With a moisture transport model calibrated to calculate wetting, the influence of cyclic ambient relative humidities on the concrete strength and stiffness could be calculated with the material model. However, in case of the time-dependent deformations of concrete, the influence of variable moisture content on creep needs to be investigated more in detail.

Chapter 8

Zusammenfassung

Die Alterung von Beton ist im Allgemeinen mit zeit- und umweltabhängigen Veränderungen der mechanischen Eigenschaften wie Kriechen, Schwinden, Elastizitätsmodul, Druck- und Zugfestigkeit verbunden. Die vorliegende Arbeit beschreibt die Entwicklung eines neuen Materialmodells für Beton zur Bewertung seiner Alterung infolge der Umweltbedingungen. Das entwickelte Modell ist in der Lage, die zeitliche Veränderung der Druck- und Zugfestigkeit, des Elastizitätsmoduls, sowie des Kriechens und Schwindens von Betonen, die erhöhten Temperaturen bis zu 100 °C bei unterschiedlichen relativen Luftfeuchten (von 30 bis 100 %) unterworfen sind, vorherzusagen.

Die Konzeption des neuen Materialmodells folgte einem rigorosen Entwicklungsprozess. Das ursprüngliche Ziel dieser Arbeit war die Entwicklung eines Modells zur Vorhersage der mechanischen Eigenschaften von Beton bei erhöhten Temperaturen bis zu 100 °C ohne Variation der Luftfeuchtigkeit. Jedoch zeigte sich im Zuge der experimentellen Untersuchungen, dass der Feuchtegehalt nicht außer Acht gelassen werden kann und im Modellansatz berücksichtigt werden muss. Die Literaturstudie zum Einfluss der Temperatur auf die mechanischen Eigenschaften von Beton zeigte deutlich, dass die Anwesenheit von Wasser in der Mikrostruktur des erhärteten Zementsteins entscheidend für das Verhalten der mechanischen Eigenschaften von Beton ist.

Darüber hinaus wurde festgestellt, dass eine erhöhte Temperatur nicht nur für die Entstehung von Schäden in der Betonmikrostruktur verantwortlich ist. Sie beeinflusst auch wesentlich die Wasserspeicherkapazität des Betons und verbessert den Austausch von Feuchtigkeit zwischen den Poren der Betonmikrostruktur und der Umgebung. Hieraus resultieren positive Auswirkungen, z.B. auch auf die Betonfestigkeit. Aus diesen Gründen musste die ursprüngliche Idee der Modellentwicklung neu bewertet und modifiziert werden. Das Modell soll nicht nur den Einfluss der Temperatur auf die mechanischen Eigenschaften von Beton vorhersagen können, sondern auch die Wechselwirkung mit dem Feuchtegelhalt zutreffend berücksichtigen.

Das Materialmodell wurde einem Entwicklungsprozess folgend aufgestellt, welcher in vier Phasen aufgeteilt werden kann. In der Literaturrecherche (erste Phase) wurden zunächst die Grundlagen der experimentellen Untersuchungen festgelegt. Die Planung und Durchführung der Experimente bildeten die zweite Phase, die der Sammlung von Daten und der Gewinnung neuer Erkenntnisse dienten. Zentrale Merkmale der Versuche waren die ortsaufgelöste Erfassung von Temperatur und Feuchte über den Querschnitt unterschiedlicher Proben aus 3 verschiedenen Betonen sowie die Entwicklung der Messtechnik, um die Proben zu konditionieren und das Kriechen von Beton bei erhöhten Temperaturen und konstanter relativen Feuchte zu messen.

Basierend auf den Ergebnissen der durchgeführten Experimente wurden anschließend in der dritten Phase weitgehend physikalisch basierte mathematische Gleichungen formuliert, mit denen die Zusammenhänge zwischen den Änderungen des Feuchtegehalts sowie der Temperatur mit Änderungen der mechanischen Eigenschaften von Beton beschrieben werden können. Schließlich wurde in der vierten Phase das Materialmodell aus den einzelnen Komponenten entwickelt. In den folgenden Abschnitten werden die wichtigsten Ergebnisse und Erkenntnisse, die während der Untersuchung erzielt wurden, zusammengefasst und abschließend die im Zuge der Ausarbeitung aufgekommenen offenen Fragen aufgeführt.

8.1 Wesentliche Ergebnisse und Erkenntnisse

Das Materialmodell wird aus drei Hauptbestandteilen gebildet: ein Modell zur Berechnung des Feuchtetransports in Beton, mathematische Gleichungen zur Berechnung der Einflüsse erhöhter Temperatur und relativer Luftfeuchte der Poren auf die Festigkeit und Steifigkeit von Beton sowie mathematische Gleichungen zur Berechnung von Kriechen und Schwinden bei erhöhten Temperaturen und unterschiedlichen Umgebungsfeuchtigkeiten. Die Entwicklung dieser Komponenten generierte Fachwissen, das in den im Folgenden aufgeführten Erkenntnissen zusammengefasst werden kann.

Erkenntnisse im Zuge der Entwicklung des Modells des Feuchtetransports

- Die Messungen zeigten, dass der Trocknungsprozess von Beton durch den Einfluss der Temperatur beschleunigt wird und mit abnehmender Porosität langsamer voranschreitet. Die Unterschiede in den Feuchtetransporteigenschaften zwischen Betonen mit niedriger Porosität und Betonen mit hoher Porosität waren ausgeprägter bei einer Temperatur von 20 °C im Vergleich zu den gleichen Betonen, die bei erhöhten Temperaturen getestet wurden. Diese Tatsache konnte durch die Entwicklung des Modells des Feuchtetransports besser verstanden werden. Bei erhöhten Temperaturen wird die Diffusivität des Betons durch die Wirkung der Aktivierungsenergie der Diffusion verbessert. Nach den experimentellen Untersuchungen und den für das Modell getroffenen Annahmen ist diese Energie um so höher, je geringer die Porosität der Betonmikrostruktur ist. Deshalb erreichten die Betone, die bei erhöhten Temperaturen getestet wurden, unabhängig von den Porositäten ihrer Mikrostrukturen ähnliche Diffusivitäten.

- Der Diffusionskoeffizient von Beton ist weder konstant noch lediglich vom Feuchtegehalt des Betons abhängig, wie es in der Literatur in der Regel angenommen wird. Die Messungen der relativen Feuchte in den Betonporen zeigten, dass die Dauer der Einstellung des Feuchtegleichgewichts mit der Umgebung unabhängig von der Umgebungsfeuchte ist. Unter Verwendung eines konstanten Diffusionskoeffizienten kann dieses Verhalten durch ein auf Diffusion basiertes Modell abgebildet werden. Allerdings kann dieses Verhalten nicht reproduziert werden, wenn ein von der relativen Feuchte abhängiger Diffusionskoeffizient verwendet wird. Weiterhin zeigten die Messungen und die weitere Berechnung mit dem Wassertransportmodell, dass die zeitliche Entwicklung der relativen Feuchte in den Betonporen während des Trocknens stark nichtlinear ist. Vor diesem Hintergrund ist die Verwendung konstanter Diffusionskoeffizienten ausgeschlossen. Um den Feuchtetransport in Beton genauer zu modellieren, war es notwendig, die Abhängigkeit von der relativen Feuchte der Poren, der relativen Feuchte der Umgebung und von den Gradienten der relativen Feuchte im Beton in die Formulierung des Diffusionskoeffizienten zu implementieren.
- Die Erhöhung der relativen Feuchte in den Betonporen infolge eines Temperaturanstiegs wurde im Rahmen der Messungen bestätigt, wie es auch nach der Definition des hygrothermischen Koeffizienten von Beton zu erwarten war. Die in der vorliegenden Arbeit für den hygrothermischen Koeffizienten von Beton vorgeschlagene Formulierung beruht auf den eigenen durchgeführten Messungen. Laut der Formulierung ist der Koeffizient um so geringer, je höher die Porosität des Betons ist. Zusätzlich ist dieser von der relativen Feuchte der Betonporen abhängig und erreicht Maximalwerte im Bereich zwischen 80 % r. F. und 90 % r. F.

Erkenntnisse im Zuge der Entwicklung der Formulierungen zur Berechnung der Festigkeit und Steifigkeit des Betons

- Die Entwicklung des experimentellen Programms erfolgte unter der Annahme, dass der Einfluss der Umgebung auf die mechanischen Betoneigenschaften mit Hilfe der Überlagerung der Auswirkungen von Temperatur und Feuchtegehalt erklärt werden kann. Unter dieser Annahme können keine zusätzlichen Effekte, die aus der Wechselwirkung zwischen Temperatur und Feuchtigkeit resultieren, berücksichtigt werden. Die Messergebnisse deuten jedoch darauf hin, dass diese Annahme nicht für Betone mit hohem Feuchtegehalt gilt. Ein dritter Effekt, nämlich die sogenannten hydrothermalen Reaktionen, ergibt sich aus der Kombination von hoher Temperatur und hohem Feuchtegehalt in der Mikrostruktur eines mit silikatischer Gesteinskörnung hergestellten Betons. Aus der Literaturrecherche war im Voraus bekannt, dass die Entwicklung von hydrothermalen Reaktionen möglicherweise eine Rolle spielen. Allerdings wird in den Literaturquellen darauf hinwiesen, dass die Einflüsse dieser Reaktionen auf die Festigkeit und Steifigkeit des Betons nur sekundär sind. Die experimentellen Ergebnisse zeigten jedoch, dass die

Betonproben, die mit hohem Feuchtegehalt in der Mikrostruktur erhöhten Temperaturen unterworfen waren, höhere Zug- und Druckfestigkeit aufwiesen. Jedoch war die Wirkung der hydrothermalen Reaktionen auf den Elastizitätsmodul der Betonproben weniger signifikant.

- Sowohl Druck- als auch Zugfestigkeit nehmen mit abnehmendem Feuchtegehalt in der Betonmikrostruktur zu. Anhand der Ergebnisse der durchgeführten Experimente von Betonen unterschiedlicher Porositäten wurde festgestellt, dass der Einfluss der Trocknung auf die Druck- und Zugfestigkeit des Betons von der Porosität der Betone unabhängig ist.

- Die Trocknung hat einen negativen Einfluss auf den Elastizitätsmodul von Beton. Die experimentellen Ergebnisse zeigten, dass die Proben, die bei einer Umgebung mit 20 °C und 100 % r. F. gelagert wurden, den höchsten Elastizitätsmodul aufwiesen. Nach einer geringen Änderung der relativen Feuchte der Betonporen wurde eine abrupte Abnahme der Steifigkeit gemessen. Mit weiterer Trocknung nahm die Steifigkeit teilweise wieder zu.

- Die Experimente zeigten, dass nach der Erhöhung der Temperatur ein sofortiger negativer Effekt hinsichtlich der Festigkeit und Steifigkeit des Materials zu verzeichnen ist. Eine zeitabhängige Erholung dieser mechanischen Eigenschaften ist immer mit einem Feuchtigkeitsverlust, weiterer Hydratation oder der Entwicklung hydrothermaler Reaktionen verbunden. Diese Prozesse werden jedoch durch die Wirkung erhöhter Temperaturen verbessert.

- Das entwickelte Materialmodell ist in der Lage, den Einfluss von erhöhten Temperaturen und Feuchtegehalt auf die Festigkeit und Steifigkeit von Beton im Bereich von 20 bis 100 °C für Feuchtigkeiten zwischen 30 und 100 % r. F. zutreffend vorherzusagen.

- Basierend auf den experimentellen Untesuchungen und unter Berücksichtigung der Formulierungen des Materialmodells können folgende Schlussfolgerungen gezogen werden:

  Die Druckfestigkeit kann bei Betonen mit geringem Feuchtegehalt (d. h. ca. 50 % r. F.) im Vergleich zu feuchten Betonen mit Feuchtigkeit nahe 100 % r. F. bis zu 32 % zunehmen. Im Fall der Zugfestigkeit beträgt diese Zunahme ungefähr 40 %.

  Erhöhte Temperaturen haben einen negativen Einfluss auf die Festigkeit und Steifigkeit des Betons. Im Bereich von 20 bis 100 °C nehmen mit steigender Temperatur im Durchschnitt die Druckfestigkeit und die Zugfestigkeit um 5 % alle 10 K ab. Im Fall des Elastizitätsmoduls beträgt diese Abnahme ca. 3 % alle 10 K.

Erkenntnisse im Zuge der Entwicklung der Formulierungen zur Berechnung des Kriechens und Schwindens von Beton

- Der Feuchtegehalt übt einen starken Einfluss auf die Kriech- und Schwinddehnungen aus. Die durchgeführten Experimente zeigten eine deutliche Abhängigkeit der Entwicklung von Kriechen und Schwinden mit dem anfänglichen Feuchtegehalt der Betonproben. Die zuvor bei niedrigeren relativen Feuchten konditionierten Proben zeigten, aufgrund des geringeren Feuchtegehaltes in ihrer Mikrostruktur, geringe Schwind- und Kriechverformungen.

- Die experimentellen Ergebnisse zeigten, dass unabhängig von Anfangsfeuchtegehalt und Temperatur der Betonproben die Änderungsrate der mittleren relativen Feuchte in eindeutiger Weise Größe und Verlauf des Trocknungsschwindens skaliert. Nach den entwickelten Formulierungen folgen das Trocknungsschwinden und die Veränderung der mittleren relativen Feuchte der gleichen Kinetik. Da die Größe des Grundkriechens durch den Feuchtegehalt des Betons beinflusst wird, ergibt sich aus den Modellformulierungen, dass auch das Trocknungskriechen der gleichen Kinetik wie die Änderung der mittleren relativen Feuchte unterliegt.

- Die Ergebnisse der Kriechversuche zeigten, dass die gesamte Kriechverformung mit steigender Temperatur zunimmt. Die Kalibrierung der Formulierungen für Kriechen erlaubte die Identifikation unterschiedlicher Effekte der Temperatur auf die Entwicklung von Grund- und Trocknungskriechen. Gemäß den Formulierungen nimmt die Komponente des Grundkriechens mit steigender Temperatur zu und die Komponente des Trocknungskriechens erreicht, trotz höherer Prozessgeschwindigkeit, niedrigere Endwerte bei erhöhten Temperaturen als bei einer Temperatur von 20 °C.

- Das entwickelte Materialmodell ist in der Lage, die zeitabhängigen Verformungen von Beton unter Berücksichtigung der Konditionierung vor und nach der Belastung für Temperaturen im Bereich von 20 bis 100 °C zu berechnen. Zu diesem Zweck war es zwingend erforderlich, im Materialmodell die Unterscheidung zwischen den Grund- und Trocknungskomponenten des Schwindens und des Kriechens zu berücksichtigen.

- Basierend auf der Auswertung der Ergebnisse der durchgeführten Experimente sowie der Ergebnisse des Materialmodells können folgende Schlussfolgerungen gezogen werden:

  Die Grundkriechverformungen sind bei 30 °C im Durchschnitt doppelt so hoch wie bei 20 °C. Das Grundkriechen ist bei einer Temperatur von 70 °C dreimal höher als bei einer Temperatur von 20 °C.

Obwohl die Endwerte der Trocknungskriech- und Trocknungsschwindverformungen stark vom Anfangsfeuchtegehalt des Betons abhängen, nehmen sie im Durchschnitt mit einer Temperaturerhöung von 10 K um 5 % ab.

8.2 Offene Fragen

Um das Ziel dieser Arbeit zu erreichen, musste über einen sehr langen Zeitraum - der diffusive Feuchtetransport im Betongefüge verläuft sehr langsam - ein erheblicher experimenteller und letztlich auch theoretischer Aufwand betrieben werden. Daher wurden einige wichtige Fragen absichtlich außerhalb des Untersuchungsrahmens gelassen. Darüber hinaus konnten nicht alle Fragen, die während der Untersuchung auftauchten, vollständig geklärt werden. Diese Aspekte werden schließlich beim Materialmodell durch die Begrenzung der Anwendbarkeit bzw. der Gültigkeit des Modells reflektiert. Die Klärung der nachfolgend aufgeführten Sachverhalte bleibt daher zukünftigen Forschungsarbeiten vorbehalten.

Alle Betonproben wurden unter Verwendung von silikatischen Gesteinskörnungen hergestellt. Der Einfluss anderer Arten von Zuschlagstoffen kann nicht durch das entwickelte Modell beschrieben werden. Für Gesteinskörnungen mit erheblich geringeren thermischen Ausdehnungskoeffizienten wie beispielsweise Kalkstein sind die entwickelten Formulierungen zur Bestimung der Temperaturänderungen mit Änderungen der Festigkeit und Steifigkeit des Betons möglicherweise nicht gültig. Mit Hilfe von zusätzlichen experimentellen Untersuchungen kann das entwickelte Modell jedoch leicht angepasst werden.

Während der gesamten Untersuchung wurde angenommen, dass das Trocknen der einzige Prozess ist, der die Kinetik des mechanischen Verhaltens der Proben, die den erhöhten Temperaturen unterworfen wurden, steuert. Es wurde jedoch festgestellt, dass hydrothermale Reaktionen bei Proben mit erhöhter Temperatur und hoher Feuchtigkeit eine Rolle spielen. In den Experimenten wurden weder der Einfluss der hydrothermalen Reaktionen auf die Druck- und Zugfestigkeit des Betons noch ihre zeitliche Entwicklung bestimmt. Aus diesen Gründen berücksichtigt das Materialmodell nicht die Wirkung hydrothermaler Reaktionen, wodurch das Modell den Einfluss erhöhter Temperaturen für Betone mit einer zu hohen Menge an Feuchtigkeit in ihrer Mikrostruktur überschätzt. Allerdings ist das entwickelte Materialmodell grundsätzlich dafür geeignet, den beschriebenen Effekt zu integrieren, wenn er zuvor experimentell eindeutig quantifiziert worden ist. Dies hätte den Rahmen der vorliegenden Arbeit jedoch gesprengt.

Basierend auf den Messungen wurde eine Formulierung zur Berechnung des Einflusses einer Temperaturänderung auf die relative Feuchte der Betonporen abgeleitet. Die für den sogenannten hygrothermischen Koeffizienten von Beton vorgeschlagene mathematische Formel beschreibt, dass die relative Feuchte der Poren mit steigender Temperatur zunimmt und mit sinkender Temperatur abnimmt. Die Formel beinhaltet jedoch nicht die zeitliche Entwicklung dieser Prozesse. Diese werden als sofortig angenommen, was für die Anwendbarkeit des Materialmodells Konsequenzen hat. Das Modell kann weder die Zeitentwicklung der Betonfestigkeit sowie der Steifigkeit noch das Kriechen und Schwinden berechnen, wenn einen Abfall der Temperatur stattfindet, sehr wohl aber die jeweiligen Endwerte. Grund hierfür ist der mehrere Tage andauernde Prozess der Reduktion der relativen Feuchte in den Betonporen. Für den Fall, dass die Temperatur zunimmt, ist die Annahme eines momentanen Inkrementes der relativen Feuchte weniger ungenau. Dieser Vorgang dauert nur wenige Stunden, sodass die Berechnung der Festigkeit und Steifigkeit nicht beeinflusst wird. Allerdings wird dadurch die Genauigkeit des Kriechmodells verringert, da die damit verbundenen Auswirkungen auf die Kriechverformung während der ersten Stunden nach der Belastung nicht richtig berücksichtigt werden können.

In den Kriechversuchen, die bei erhöhten Temperaturen durchgeführt wurden, erfolgte die Belastung der Proben nach Erreichen der Versuchstemperatur. Obwohl das Materialmodell in der Lage ist, den Einfluss der Erwärmung während der Belastung auf die Kriechverformungen zu berechnen, wurden solche Fälle nicht untersucht. Hierzu lagen keine eigene Experimente zur Verifizierung der Prognosen des Modells vor. Ein erster orientierender Vergleich mit ensprechenden Ergebnissen aus der Literatur belegte, dass das Modell auch den Effekt des transienten Kriechens zutreffend wiedergibt.

Der Prozess der Befeuchtung von Beton wurde nur theoretisch angegangen. Um die für den Diffusionskoeffizienten vorgeschlagene Definition im Falle der Befeuchtung zu verifizieren, müssen entsprechende (experimentelle) Messungen durchgeführt werden. Mit einem zur Berechnung der Auswirkungen der Befeuchtung kalibrierten Wassertransportmodell könnte der Einfluss von zyklischen relativen Luftfeuchten auf die Festigkeit und Steifigkeit von Beton mit dem entwickelten Materialmodell beschrieben werden. Bei den zeitabhängigen Verformungen von Beton muss jedoch der Einfluss des variablen Feuchtegehalts auf das Kriechen näher untersucht werden.

Bibliography

[1] ADOLPHS, J. Surface energies of hardened cement paste depending on relative humidity. *Materials and Structures 38*, 4 (2005), 443 – 448.

[2] ADOLPHS, J., SETZER, M. J., AND HEINE, P. Changes in pore structure and mercury contact angle of hardened cement paste depending on relative humidity. *Materials and Structures 35*, 8 (2002), 477 – 486.

[3] ASSARSSON, G. O. Hydrothermal reactions of calcium hydroxide - quartz at 120 - 220 °C. *The Journal of Physical Chemistry 64*, 3 (1960), 328 – 331.

[4] ATKINS, P. W., AND DE PAULA, J. *Atkins physical chemistry*. Oxford Univ. Press, Oxford, 2014.

[5] BAŽANT, Z. Constitutive equation for concrete creep and shrinkage based on thermodynamics of multiphase systems. *Matériaux et Construction 3*, 1 (1970), 3 – 36.

[6] BAŽANT, Z. Delayed thermal dilatations of cement paste and concrete due to mass transport. *Nuclear Engineering and Design 14*, 2 (1970), 308 – 318.

[7] BAŽANT, Z. Thermodynamics of hindered adsorption and its implications for hardened cement paste and concrete. *Cement and Concrete Research 2*, 1 (1972), 1 – 16.

[8] BAŽANT, Z. Thermodynamics of interacting continua with surfaces and creep analysis of concrete structures. *Nuclear Engineering and Design 20*, 2 (1972), 477 – 505.

[9] BAŽANT, Z. Theory of creep and shrinkage in concrete structures: A precis of recent developments. *Mechanics today 2* (1975), 1 – 93.

[10] BAŽANT, Z., AND NAJJAR, L. Drying of concrete as a nonlinear diffusion problem. *Cement and Concrete Research 1*, 5 (1971), 461 – 473.

[11] BAŽANT, Z., AND NAJJAR, L. Nonlinear water diffusion in nonsaturated concrete. *Matériaux et Construction 5*, 1 (1972), 3 – 20.

[12] BAŽANT, Z. P., AND BAZANT, M. Z. Theory of sorption hysteresis in nanoporous solids: Part I: Snap-through instabilities. *Journal of the Mechanics and Physics of Solids 60*, 9 (2012), 1644 – 1659.

[13] BAŽANT, Z. P., AND WITTMANN, F. H. *Creep and shrinkage in concrete structures*. Wiley New York, 1982.

[14] BABUŠKIN, V. I. *Thermodynamik der Silikate*, 2. ed. Verl. f. Bauwesen, Berlin, 1966.

[15] BANGHAM, D. H., AND FAKHOURY, N. CLXXV.-The translation motion of molecules in the adsorbed phase on solids. *J. Chem. Soc.* (1931), 1324–1333.

[16] BAROGHEL-BOUNY, V. Water vapour sorption experiments on hardened cementitious materials: Part I: Essential tool for analysis of hygral behaviour and its relation to pore structure. *Cement and Concrete Research 37*, 3 (2007), 414 – 437.

[17] BARTON, T. J., BULL, L. M., KLEMPERER, W. G., LOY, D. A., MCENANEY, B., MISONO, M., MONSON, P. A., PEZ, G., SCHERER, G. W., VARTULI, J. C., AND YAGHI, O. M. Tailored porous materials. *Chemistry of Materials 11*, 10 (1999), 2633 – 2656.

[18] BAZANT, M. Z., AND BAŽANT, Z. P. Theory of sorption hysteresis in nanoporous solids: Part II: Molecular condensation. *Journal of the Mechanics and Physics of Solids 60*, 9 (2012), 1660 – 1675.

[19] BEDNAR, T. *Beurteilung des feuchte- und wärmetechnischen Verhaltens von Bauteilen und Gebäuden - Weiterentwicklung der Meß- und Rechenverfahren*. 2000. Dissertation, Technical University Vienna.

[20] BEHNOOD, A., AND ZIARI, H. Compressive strength of high-strength concretes at temperatures up to 300 °C. In *Keep Concrete Attractive Proc. of the fib symposium* (2003), vol. 2, pp. 1203 – 1208.

[21] BENTZ, D. P., SNYDER, K. A., AND STUTZMAN, P. E. Hydration of Portland cement: The effects of curing conditions. *Proceedings of the 10th International Congress on the Chemistry of Cement, Vol. 2, Gothenburg, Sweden* (1997).

[22] BINGÖL, A. F., AND GÜL, R. Effect of elevated temperatures and cooling regimes on normal strength concrete. *Fire and Materials 33*, 2 (2009), 79 – 88.

[23] BONNAUD, P. A., COASNE, B., AND PELLENQ, R. J.-M. Molecular simulation of water confined in nanoporous silica. *Journal of Physics: Condensed Matter 22*, 28 (2010), 1 – 15.

[24] BONNAUD, P. A., JI, Q., COASNE, B., PELLENQ, R. J.-M., AND VAN VLIET, K. J. Thermodynamics of water confined in porous calcium-silicate-hydrates. *Langmuir 28*, 31 (2012), 11422 – 11432.

[25] BONNAUD, P. A., JI, Q., AND VAN VLIET, K. J. Effects of elevated temperature on the structure and properties of calcium–silicate–hydrate gels: the role of confined water. *Soft Matter 9*, 28 (2013), 6418 – 6429.

[26] BONZEL, JUSTUS; KADLEČEK, V. Einfluß der Nachbehandlung und des Feuchtigkeitszustands auf die Zugfestigkeit des Betons. *Beton 20*, 7 (1970), 303 – 309.

[27] BROCHARD, L., VANDAMME, M., AND PELLENQ, R.-M. Poromechanics of microporous media. *Journal of the Mechanics and Physics of Solids 60*, 4 (2012), 606 – 622.

[28] BROUWERS, H. The work of Powers and Brownyard revisited: Part 1. *Cement and Concrete Research 34*, 9 (2004), 1697 – 1716.

[29] BROUWERS, H. The work of Powers and Brownyard revisited: Part 2. *Cement and Concrete Research 35*, 10 (2005), 1922 – 1936.

[30] BROWN, N. H., AND HOPE, B. B. The creep of hydrated cement paste. *Cement and Concrete Research 6*, 4 (1976), 475 – 485.

[31] BROWNE, R. Properties of concrete in reactor vessels. In *Prestressed Concrete Pressure Vessel Conference* (1967), vol. Group C, pp. 131 – 151.

[32] BROWNE, R. D., AND BLUNDELL, R. The influence of loading age and temperature on the long term creep behaviour of concrete in a sealed, moisture stable, state. *Matériaux et Construction 2*, 2 (1969), 133–143.

[33] BRUNAUER, S. *The adsorption of gases and vapors*, vol. 1: Physical adsorption. Princeton Univ. Press, Princeton, 1945.

[34] BRUNAUER, S., EMMETT, P. H., AND TELLER, E. Adsorption of gases in multimolecular layers. *Journal of the American Chemical Society 60*, 2 (1938), 309 – 319.

[35] BUDELMANN, H. *Verhalten von Beton bei mäßig erhöhten Betriebstemperaturen*. Deutscher Ausschuß für Stahlbeton; 404. Beuth, Berlin, 1989.

[36] CAMPBELL-ALLEN, D., AND DESAI, P. The influence of aggregate on the behaviour of concrete at elevated temperatures. *Nuclear Engineering and Design 6*, 1 (1967), 65 – 77.

[37] CASTILLO, C., AND DURRANI, A. Effect of transient high temperature on high-strength concrete. *ACI Materials journal 87*, 1 (1990).

[38] CHENG, F.-P., KODUR, V., AND WANG, T.-C. Stress-strain curves for high strength concrete at elevated temperatures. *Journal of Materials in Civil Engineering 16*, 1 (2004), 84 – 90.

[39] CLAISSE, P. A. *Transport Properties of Concrete: Measurements and Applications*. Woodhead Publishing Series in Civil and Structural Engineering; 53. Woodhead Publ., Elsevier, Amsterdam, 2014.

[40] CZERNIN, W. *Zementchemie für Bauingenieure*. Bauverlag, 1977.

[41] DAHMS, J. Einfluß der Eigenfeuchtigkeit auf die Druckfestigkeit des Betons. *Beton 18*, 9 (1968), 361 – 365.

[42] DAVIS, R., DAVIS, H., AND HAMILTON, J. Plastic flow of concrete under sustained stress. In *ASTM Proceeding* (1934), vol. 34, pp. 354 – 386.

[43] DE VRIES, D. A. Simultaneous transfer of heat and moisture in porous media. *Eos, Transactions American Geophysical Union 39*, 5 (1958), 909–916.

[44] DERJAGUIN, B. V. *Theory of stability of colloids and thin films*. Consultants Bureau, New York, 1989.

[45] DOMONE, P. L. J., AND ILLSTON, J., Eds. *Construction materials: Their nature and behaviour*, 4th ed. Spon Press/Taylor & Francis, London, 2010.

[46] DUHR, S., AND BRAUN, D. Why molecules move along a temperature gradient. *Proceedings of the National Academy of Sciences of the United States of America 103*, 52 (2006), 19678 – 19682.

[47] EHM, C., HINRICHSMEYER, K., AND DIEDERICHS, U. Mechanical and physical properties of flyash concrete after hydrothermal storage. *Cement and Concrete Research 17*, 6 (1987), 968 – 976.

[48] ČERNÝ, R., AND ROVNANÍKOVÁ, P. *Transport processes in concrete*. Spon, London, 2002.

[49] ESPINOSA, R. M. *Sorptionsisothermen von Zementstein und Mörtel*. PhD thesis, Herdecke, 2005. Hamburg, Techn. Univ., Diss., 2004.

[50] ESPINOSA, R. M., AND FRANKE, L. Influence of the age and drying process on pore structure and sorption isotherms of hardened cement paste. *Cement and Concrete Research 36*, 10 (2006), 1969 – 1984.

[51] FELDMAN, R. Mechanism of creep of hydrated portland cement paste. *Cement and Concrete Research 2*, 5 (1972), 521 – 540.

[52] FELDMAN, R. F. Sorption and length-change scanning isotherms of methanol and water on hydrated Portland cement. In *Proceedings of the 5th International Symposium on the Chemistry of Cement* (1970), no. 3, pp. 53 – 66.

[53] FELDMAN, R. F., AND SEREDA, P. J. Sorption of water on compacts of bottle-hydrated cement. I. The sorption and length-change isotherms. *Journal of Applied Chemistry 14*, 2 (1964), 87 – 93.

[54] FELDMAN, R. F., AND SEREDA, P. J. A model for hydrated Portland cement paste as deduced from sorption-length change and mechanical properties. *Matériaux et construction 1*, 6 (1968), 509 – 520.

[55] FELDMAN, R. F., AND SEREDA, P. J. A new model for hydrated Portland cement and its practical implications. *Engineering Journal 53*, 8/9 (1970), 53 – 59.

[56] FOURNIER, R. O., AND ROWE, J. J. The solubility of amorphous silica in water at high temperatures and high pressures. *American Mineralogist 62* (1977), 1052 – 1056.

[57] FUNK, M. Hysteretic moisture properties of porous materials: Part I: Thermodynamics. *Journal of Building Physics 38*, 1 (2014), 6–49.

[58] FUNK, M., AND WAKILI, K. G. Driving potentials of heat and mass transport in porous building materials: A comparison between general linear, thermodynamic and micromechanical derivation schemes. *Transport in Porous Media 72*, 3 (2008), 273–294.

[59] FURUMURA, F., AVE, T., SHINOHARA, Y., AND ABE, T. Mechanical properties of high strength concrete at high temperatures. In *Proceedings of the Fourth Weimar Workshop on High Performance Concrete* (1995), pp. 237 – 252.

[60] GARRECHT, H. *Porenstrukturmodelle für den Feuchtehaushalt von Baustoffen mit und ohne Salzbefrachtung und rechnerische Anwendung auf Mauerwerk.* PhD thesis, Karlsruhe, 1992.

[61] GEIKER, M. *Studies of Portland Cement Hydration by Measurements of Chemical Shrinkage and a Systematic Evaluation of Hydration Curves by Means of the Dispersion Model.* Ph.D. Thesis. Technical University of Denmark, 1983.

[62] GERTIS, K. M., Ed. *Hygrische Transportphänomenenomene in Baustoffen.* Deutscher Ausschuß für Stahlbeton; 258. Ernst, Berlin, 1976.

[63] GIBBS, J. W. *The Scientific Papers of J. Willard Gibbs.* Woodbridge, CT: Ox Bow Press, [1906], 1993.

[64] GRÜBL, P., WEIGLER, H., AND KARL, S. *Beton: Arten, Herstellung, Eigenschaften*, 2. ed. Handbuch für Beton-, Stahlbeton- und Spannbetonbau. Ernst, Berlin, 2001.

[65] GRUNEWALD, J. *Diffusiver und konvektiver Stoff- und Energietransport in kapillarporösen Baustoffen.* Dresdner bauklimatische Hefte. TU Dresden, 1997.

[66] GUGGENHEIM, E. A. *Thermodynamics: an advanced treatment for chemists and physicists*, 5. ed. North-Holland, Amsterdam, 1967.

[67] GUNDLACH, H. *Dampfgehärtete Baustoffe*. Bauverlag, Wiesbaden, Berlin, 1973.

[68] GUNNARSSON, I., AND ARNÓRSSON, S. Amorphous silica solubility and the thermodynamic properties of $H_4SiO°_4$ in the range of 0° to 350°C at P_{sat}. *Geochimica et Cosmochimica Acta 64*, 13 (2000), 2295 – 2307.

[69] GUO, J., AND WALDRON, P. Deterioration of PCPV concrete. *Nuclear Engineering and Design 198*, 3 (2000), 211 – 226.

[70] HANNANT, D. J. *Strain behaviour of concrete up to 95 °C under compressive stresses*. 1967, pp. 171–191.

[71] HANSEN, T. Physical structure of hardened cement paste. A classical approach. *Materials and Structures 19*, 6 (1986), 423 – 436.

[72] HANSEN, T. C., AND MATTOCK, A. H. Influence of size and shape of member on the shrinkage and creep of concrete. *Journal Proceedings ACI 63*, 2 (1966), 267 – 290.

[73] HANSON, J. Effects of curing and drying enviroments on splitting tensile strength of concrete. In *ACI Journal Proceedings* (1968), vol. 65, ACI, pp. 535 – 543.

[74] HARMATHY, T. Moisture sorption of building materials. *NRC Canada, Div. Building. Res. Techn. paper Nr. 242* (1967).

[75] HASSANIZADEH, S., AND GRAY, W. G. Mechanics and thermodynamics of multiphase flow in porous media including interphase boundaries. *Advances in Water Resources 13*, 4 (1990), 169 – 186.

[76] HOFF, G. C., BILODEAU, A., AND MALHOTRA, V. M. Elevated temperature effects on HSC residual strength. *Concrete International 22*, 4 (2000), 41 – 47.

[77] HUNDT, J., AND KANTELBERG, H. *Sorptionsuntersuchungen an Zementstein, Zementmörtel und Beton*. Deutscher Ausschuß für Stahlbeton; 297. Ernst & Sohn, Berlin, 1978.

[78] ISHIDA, T., MAEKAWA, K., AND KISHI, T. Enhanced modeling of moisture equilibrium and transport in cementitious materials under arbitrary temperature and relative humidity history. *Cement and Concrete Research 37*, 4 (2007), 565 – 578.

[79] KAVIANY, M. *Principles of heat transfer in porous media*, 2. ed. Mechanical engineering series. Springer, New York, 1995.

[80] KIESSL, K. *Kapillarer und dampfförmiger Feuchtetransport in mehrschichtigen Bauteilen : rechnerische Erfassung u. bauphysikalische Anwendung*. PhD thesis, Essen, Gesamthochsch., 1983.

[81] KIESSL, K., AND GERTIS, K. *Nichtisothermer Feuchtetransport in dickwandigen Betonteilen von Reaktordruckbehältern*. Deutscher Ausschuß für Stahlbeton; 280. Ernst, Berlin, 1977.

[82] KIESSL, K., AND GERTIS, K. *Feuchtetransport in Baustoffen: eine Literaturauswertung zur rechnerischen Erfassung hygrischer Transportphänomene*. Forschungsberichte aus dem Fachbereich Bauwesen / Universität Essen, Gesamthochschule; 13. Univ., FB Bauwesen, Essen, 1980.

[83] KIELSGAARD-HANSEN, K. *Sorption isotherms: a catalogue*. Teknisk rapport / Laboratoriet for Bygningsmaterialer, Danmarks Tekniske Höjskole; 162. Technical University of Denmark, København, 1986.

[84] KLOPFER, H. *Wassertransport durch Diffusion in Feststoffen: insbesondere in Baustoffen, Kunststoffen, Beschichtungen*. Bauverl., Wiesbaden, 1974.

[85] KÜNZEL, H. M. *Simultaneous heat and moisture transport in building components: one- and two-dimensional calculation using simple parameters*. IRB-Verl., Stuttgart, 1995.

[86] KONDO, R. Kinetic study on hydrothermal reaction between lime and silica. In *Proc. first Int. Symp. on Autoclaved Calcium Silicate Buildings Products* (1967), pp. 92–97.

[87] KOOI, J. V. D. *Moisture transport in cellular concrete roofs*. PhD thesis, Technische Hogeschool Eindhoven, 1971.

[88] KORDINA, K., AND BUDELMANN, H. *Zum Festigkeits- und Verformungsverhalten von Beton bei instationärem Umgebungsklima*. Literatursichtung zum Kriechen von Beton bei erhöhter Temperatur bis 100 °C. Institut f. Baustoffe, Massivbau und Brandschutz, Braunschweig, 1984.

[89] KOTTAS, R., SEEBERGER, J., AND HILSDORF, H. Strength characteristics of concrete in the temperature range of 20 °C to 200 °C. In *5th SMIRT Int. Conference, Berlin* (1979). Paper H1/2.

[90] KRISCHER, O. *Die wissenschaftlichen Grundlagen der Trocknungstechnik*, 3th ed. Springer, Berlin, 1978.

[91] KÜTTNER, C. H., AND EHLERT, G. Experimental investigations of creep of concrete at temperatures up to 130 °C and boundary moisture conditions. *Wiss. Hochsch. Archit. Bauwes. B. Weimar*, 38 (1992), 211 – 218.

[92] LANGMUIR, I. The adsorption of gases on plane surfaces of glass, mica and platinum. *Journal of the American Chemical Society 40*, 9 (1918), 1361–1403.

[93] LUIKOV, A. V. *Heat and mass transfer in capillary-porous bodies*, 1st engl. ed. Pergamon Pr., Oxford, 1966.

[94] MINDESS, S., AND YOUNG, J. F. *Concrete*. Prentice Hall, Englewood Cliffs, New Jersey, 1981.

[95] MITZEL, A., AND KLAPOC, M. On the superposition of shrinkage and creep deformations. *Building Science 2*, 3 (1967), 267 – 271.

[96] MÜLLER, H. S. *Zur Vorhersage des Kriechens von Konstruktionsbeton*. PhD thesis, Karlsruhe University, 1986.

[97] MÜLLER, H. S., AND ACOSTA, F. Time dependent effects of structural concrete: Basics for constitutive modelling towards the next generation of EN Eurocode 2. In *Massivbau im Wandel, Festschrift zum 60. Geburtstag von Josef Hegger* (2014), Ernst & Sohn, Berlin, pp. 395 – 413.

[98] MÜLLER, H. S., AND KVITSEL, V. Kriechen und Schwinden von Beton. Grundlagen der neuen DIN 1045 und Ansätze für die Praxis. *Beton- und Stahlbetonbau 97*, 1 (2002), 8 – 19.

[99] MONFORE, G. A small probe-type gage for measuring relative humidity. *The Journal of PCA Research and Development Laboratories 5*, 2 (1963), 41 – 47.

[100] MOREY, G., AND HESSELGESSER, J. Solubility of quartz and some other substances in superheated steam at high pressures. *Trans. ASME (Am. Soc. Mech. Eng.); (United States)* (Jan 1951).

[101] MOREY, G., AND HESSELGESSER, J. Solubility of some minerals in superheated steam at high pressures. *Econ. Geol.; (United States) 46:8* (Jan 1951).

[102] NASSER, K. W., AND NEVILLE, A. M. Creep of old concrete at normal and elevated temperatures. *Journal Proceedings ACI 64*, 2 (1967), 97 – 103.

[103] NAUS, D. J. *A compilation of elevated temperature concrete material property data and information for use in assessments of nuclear power plant reinforced concrete structures*. U.S. Nuclear Regulatory Commission, Office of Nuclear Regulatory Research, 2010.

[104] NAUS, D. J. *The Effect of Elevated Temperature on Concrete Materials and Structures - a Literature Review.* U.S. Nuclear Regulatory Commission, Office of Nuclear Regulatory Research, 2006.

[105] NEVILLE, A. M. Creep of concrete as a function of its cement paste content. *Magazine of Concrete Research 16*, 46 (1964), 21–30.

[106] NEVILLE, A. M. *Properties of concrete*, 5th ed. Pearson, Harlow, England, 2011.

[107] NEVILLE, A. M., DILGER, W., AND BROOKS, J. J. *Creep of plain and structural concrete*. Construction Pr., London, 1983.

[108] NEVILLE, A. M., AND MEYERS, B. L. Creep of concrete: Influencing factors and prediction. In *Symposium on Creep of Concrete* (1964), ACI, Ed., vol. 9 of *Special Publication*, pp. 1 – 33.

[109] NIELSEN, C., AND BIÉANIĆ, N. Residual fracture energy of high-performance and normal concrete subject to high temperatures. *Materials and Structures 36*, 8 (2003), 515 – 521.

[110] PHAN, L., LAWSON, J., AND DAVIS, F. Effects of elevated temperature exposure on heating characteristics, spalling, and residual properties of high performance concrete. *Materials and Structures 34*, 2 (2001), 83 – 91.

[111] PHAN, L. T., AND CARINO, N. J. Effects of test conditions and mixture proportions on behavior of high-strength concrete exposed to high temperatures. *ACI Materials Journal 99*, 1 (2002).

[112] PIHLAJAVAARA, S. A review of some of the main results of a research on the ageing phenomena of concrete: Effect of moisture conditions on strength, shrinkage and creep of mature concrete. *Cement and Concrete Research 4*, 5 (1974), 761 – 771.

[113] PINEAUD, A., REMOND, S., CABRILLAC, R., MENOU, A., BOUSSA, H., AND MOUNAJED, G. Etude expérimentale de l'influence des hautes températures sur l'énergie de fissuration des BHP. In *XXIemes Rencontres universitaires de Genie Civil 2003* (2003), pp. 257 – 264.

[114] POLIVKA, M., PIRTZ, D., AND ADAMS, R. F. Studies of creep in mass concrete. In *Symposium of Mass concrete* (1963), ACI, Ed., vol. 6 of *Special Publication*.

[115] POWERS, T. C. Absorption of water by Portland cement paste during the hardening process. *Industrial & Engineering Chemistry 27*, 7 (1935), 790 – 794.

[116] POWERS, T. C. A discussion of cement hydration in relation to the curing of concrete. *Proceedings of the Highway Research Board 27* (1947), 178 – 189.

[117] POWERS, T. C. Structure and physical properties of hardened Portland cement paste. *Journal of the American Ceramic Society 41*, 1 (1958), 1 – 6.

[118] POWERS, T. C. Mechanisms of shrinkage and reversible creep of hardened cement paste. In *International Conference on the Structure of Concrete* (1965), Cement and Concrete Association, London, pp. 319 – 344.

[119] POWERS, T. C. The thermodynamics of volume change and creep. *Matériaux et Construction 1*, 6 (1968), 487 – 507.

[120] POWERS, T. C., AND BROWNYARD, T. L. *Studies of the physical properties of hardened Portland cement paste*. Bulletin / Research Laboratories, Portland Cement Association; 22. Portland Cement Association, Research Laboratories, Chicago, Ill., 1948.

[121] POYET, S. Experimental investigation of the effect of temperature on the first desorption isotherm of concrete. *Cement and Concrete Research 39*, 11 (2009), 1052 – 1059.

[122] RAJU, M. P., AND RAO, A. J. Effect of temperature on residual compressive strength of flyash concrete. *The Indian Concrete Journal*, May (2001), 347 – 350.

[123] RAJU, M. P., RAO, K. S., AND RAJU, P. Compressive strength of heated high-strength concrete. *Magazine of Concrete Research 59*, 2 (2007), 79 – 85.

[124] RAO, K. S., RAJU, M. P., AND RAJU, P. Effect of elevated temperature exposure on split tensile strength of high strength concrete. In *Keep Concrete Attractive Proc. of the fib symposium* (2003), vol. 2, pp. 1187 – 1192.

[125] RHODES, J. A. Prediction of creep, shrinkage, and temperature effects in concrete structures. Tech. rep., American Concrete Institute. Committee on Creep and Shrinkage of Concrete, Detroit, Mich., 1992.

[126] ROSE, D. A. Water movement in porous materials: Part 1 - Isothermal vapour transfer. *British Journal of Applied Physics 14*, 5 (1963), 256.

[127] ROSE, D. A. Water movement in porous materials: Part 2 - The separation of the components of water movement. *British Journal of Applied Physics 14*, 8 (1963), 491.

[128] ROSTÁSY, F., AND BUDELMANN, H. Creep of concrete with variable moisture content at elevated temperature up to 90 °C. *4th RILEM Int. Symp. on Creep and Shrinkage of Concrete; Mathematical Modeling* (1986), 581 – 590.

[129] ROSTÁSY, F., AND BUDELMANN, H. Strength and deformation of concrete with variable content of moisture at elevated temperature up to 90 °C. *Cement and Concrete Research 16*, 3 (1986), 353 – 362.

[130] ROSTÁSY, F. S., AND BUDELMANN, H. *Prüftechnologie des Betonkriechens bei erhöhter Temperatur und veränderlicher Feuchte im instationären Zustand*. Forschungsbericht, TU Braunschweig. Institut f. Baustoffe, Massivbau und Brandschutz, Braunschweig, 1985.

[131] ROSTÁSY, F. S., EHM, C., HINRICHSMEYER, K., AND DIEDERICHS, U. Untersuchungen zum Einfluß hoher Betriebstemperaturen auf den Korrosionsschutz in Betonbauwerken der Energietechnik. Abschlussbericht, Braunschweig: Institut für Baustoffe, Massivbau und Brandschutz IBMB, 1985.

[132] ROSTÁSY, F.S.; BUDELMANN, H. Prüftechnologie des Betonkriechens bei erhöhter Temperatur und veränderlicher Feuchte im instationären Zustand. *Kurzberichte aus der Bauforschung 28* (1987), 591 – 594.

[133] RUETZ, W. *Das Kriechen des Zementsteins im Beton und seine Beeinflussung durch gleichzeitiges Schwinden: experimentelle Studien zur Klärung d. Mechanismus*. Deutscher Ausschuß für Stahlbeton; 183. Ernst, Berlin, 1966.

[134] SARSHAR, R., AND KHOURY, G. Material and environmental factors influencing the compressive strength of unsealed cement paste and concrete at high temperatures. *Magazine of concrete research 45*, 162 (1993), 51 – 61.

[135] SAVVA, A., MANITA, P., AND SIDERIS, K. Influence of elevated temperatures on the mechanical properties of blended cement concretes prepared with limestone and siliceous aggregates. *Cement and Concrete Composites 27*, 2 (2005), 239 – 248.

[136] SCHNEIDER, U. *Verhalten von Beton bei hohen Temperaturen = Behaviour of concrete at high temperatures*. Deutscher Ausschuss für Stahlbeton; 337. Ernst in Komm., Berlin, 1982.

[137] SCHWESINGER, P., EHLERT, G., AND WÖLFEL, R. Creep of concrete at elevated temperatures and boundary conditions of moisture. *Cement and Concrete Research 17*, 2 (1987), 263 – 272.

[138] SEEBERGER, J., KROPP, J., AND HILSDORF, H. K. *Festigkeitsverhalten und Strukturänderungen von Beton bei Temperaturbeanspruchung bis 250 °C*. Deutscher Ausschuß für Stahlbeton; 360. Ernst, Berlin, 1985.

[139] SEKI, S., AND KAWASUMI, M. Creep of concrete at elevated temperatures. *Concrete of Nuclear Reactors 1* (1972), 591 – 638. ACI Special Publication SP-34, Detroit.

[140] SETZER, M. Basis of testing the freeze-thaw resistance—surface and internal deterioration. In *Frost resistance of concrete, Proceedings of the International RILEM Workshop* (1997), E & FN Spon, London, pp. 157 – 173.

[141] SETZER, M. The solid-liquid gel-system of hardened cement paste / Das Porenwasser-Feststoff System des Zementgels (SLGS System). *Restoration of Buildings and Monuments 14*, 4 (2006), 259 – 270.

[142] SETZER, M. J. *Oberflächenenergie und mechanische Eigenschaften des Zementsteins*. PhD thesis, München, Techn. Univ., Fak. f. Bauwesen, 1972.

[143] SETZER, M. J. *Einfluß des Wassergehalts auf die Eigenschaften des erhärteten Betons*. Deutscher Ausschuß für Stahlbeton; 280. Ernst & Sohn, Berlin, 1977.

[144] SPLITTGERBER, H. Spaltdruck zwischen Festkorpern und Auswirkungen auf Probleme in der Technik. *Cement and Concrete Research 6*, 1 (1976), 29 – 35.

[145] SPLITTGERBER, H., AND WITTMANN, F. Einfluss adsorbierter Wasserfilme auf die van der Waals Kraft zwischen Quarzglasoberflächen. *Surface Science 41*, 2 (1974), 504 – 514.

[146] STARK, J., MÖSER, B., AND ECKART, A. Neue Ansätze zur Zementhydratation (Teil 1). *ZKG International 54*, 1 (2001), 52 – 60.

[147] STARK, J., MÖSER, B., AND ECKART, A. Neue Ansätze zur Zementhydratation (Teil 2). *ZKG International 54*, 2 (2001), 114 – 119.

[148] TAYLOR, H. F. W. *Cement chemistry*, 2. ed. Thomas Telford, London, 1997.

[149] TAZAWA, E., MIYAZAWA, S., AND KASAI, T. Chemical shrinkage and autogenous shrinkage of hydrating cement paste. *Cement and Concrete Research 25*, 2 (1995), 288 – 292.

[150] TROXELL, G., RAPHAEL, J., AND DAVIS, R. Long-time creep and shrinkage tests of plain and reinforced concrete. In *ASTM Proceeding* (1958), vol. 58, pp. 1101 – 1120.

[151] WHITAKER, S. Flow in porous media I: A theoretical derivation of Darcy's law. *Transport in Porous Media 1*, 1 (1986), 3–25.

[152] WITTMANN, F. H. *Grundlage eines Modells zur Beschreibung charakteristischer Eigenschaften des Betons*. Deutscher Ausschuß für Stahlbeton; 290. Ernst, Berlin, 1977.

[153] WU, B., SU, X.-P., LI, H., AND YUAN, J. Effect of high temperature on residual mechanical properties of confined and unconfined high-strength concrete. *ACI Materials Journal 99*, 4 (2002).

[154] XI, Y., BAŽANT, Z. P., AND JENNINGS, H. M. Moisture diffusion in cementitious materials - Adsorption isotherms. *Advanced Cement Based Materials 1*, 6 (1994), 248 – 257.

[155] XI, Y., BAŽANT, Z. P., MOLINA, L., AND JENNINGS, H. M. Moisture diffusion in cementitious materials - Moisture capacity and diffusivity. *Advanced Cement Based Materials 1*, 6 (1994), 258 – 266.

[156] XIAO, J., AND FALKNER, H. On residual strength of high-performance concrete with and without polypropylene fibres at elevated temperatures. *Fire Safety Journal 41*, 2 (2006), 115 – 121.

[157] YORK, G., KENNEDY, T., AND PERRY, E. *Experimental investigation of creep in concrete subjected to multiaxial compressive stresses and elevated temperatures.* Research report. Department of Civil Engineering, University of Texas at Austin, 1970.

[158] ZHANG, B., BICANIC, N., PEARCE, C., AND BALABANIC, G. Residual fracture properties of normal- and high-strength concrete subject to elevated temperatures. *Magazine of concrete research 52*, 2 (2000), 123 – 136.

[159] ZOLDNERS, N. Effect of high temperatures on concretes incorporating different aggregates. Tech. rep., Canada. Dept. of Mines and Technical Surveys. Mines Branch, 1960.

Codes and Guidelines

[N1] DIN 1045-2: Tragwerke aus Beton, Stahlbeton und Spannbeton - Teil 2: Beton - Festlegung, Eigenschaften, Herstellung und Konformität - Anwendungsregel zu DIN EN 206-1. Beuth Verlag, Berlin, 2008.

[N2] DIN 1048-5: Prüfverfahren für Beton - Teil 5: Festbeton, gesondert hergestellte Probekörper. Beuth Verlag, Berlin 1991.

[N3] DIN 66133: Bestimmung der Porenvolumenverteilung und der spezifischen Oberfläche von Festoffen durch Quecksilberintrusion. Beuth Verlag, Berlin, 1993.

[N4] DIN EN 12350-5: Prüfung von Frischbeton - Teil 5: Ausbreitmaß. Beuth Verlag, Berlin, 2000.

[N5] DIN EN 12350-6: Prüfung von Frischbeton - Teil 6: Frischbetonrohdichte. Beuth Verlag, Berlin 2000.

[N6] DIN EN 12350-7: Prüfung von Frischbeton - Teil 7: Luftgehalte - Druckverfahren. Beuth Verlag, Berlin 2000.

[N7] DIN EN 12390-2: Prüfung von Festbeton - Teil 2: Herstellung und Lagerung von Probekörpern für Festigkeitsprüfungen. Beuth Verlag, Berlin 2001.

[N8] DIN EN 12390-3: Prüfung von Festbeton - Teil 2: Druckfestigkeit von Proberkörpern. Beuth Verlag, Berlin 2002.

[N9] DIN EN 206-1: Beton - Teil 1: Festlegung, Eigenschaften, Herstellung und Konformität. Beuth Verlag, Berlin 2001.

[N10] DIN EN ISO 12570: Wärme- und feuchtetechnisches Verhalten von Baustoffen und Bauprodukten - Bestimmung des Feuchtegehaltes durch Trocknen bei erhöhter Temperatur. Beuth Verlag, Berlin 2013.

[N11] DIN EN ISO 12571: Wärme- und feuchtetechnisches Verhalten von Baustoffe und Bauprodukten - Bestimmung der hygroskopischen Sorptionseigenschaften. Beuth Verlag, Berlin 2013.

[N12] DIN ISO 9277: Bestimmung der spezifischen Oberfläche von Festkörpern mittels Gasadsorption - BET-Verfahren, Beuth Verlag, Berlin 2014.

[N13] DAfStb-Booklet 422: Prüfung von Beton. Empfehlungen und Hinwise als Ergänzung zu DIN 1048. Deutscher Ausschuss für Stahlbeton (Edit.), Beuth Verlag, Berlin 1991.

[N14] CEB-FIP Model Code 1990: Euro-International Committee for Concrete (CEB) and International Federation of Prestressing (FIP) (Edit.), Thomas Telford, London 1993.

[N15] fib Model Code for Concrete Structures 2010: International Federation for Structural Concrete (fib) (Edit.), Ernst & Sohn, Berlin 2013.

[N16] EN 1992-2: Eurocode 2 - Design of concrete structures - Concrete bridges - Design and detailing rules, October 2005.

[N17] IUPAC Solubility Data Series: Alkaline earth hydroxides in water and aqueous solutions, Vol. 52. International Union of Pure and Applied Chemistry (Edit.), Pergamon Press, 1992.

[N18] IUPAC Manual of symbols and terminology for physicochemical quantities and units - Appendix II: Definitions, Terminology and Symbols in Colloid and Surface Chemistry, Part II: Heterogeneous Catalysis, Vol. 46. International Union of Pure and Applied Chemistry (Edit.), Pergamon Press, 1976.

[N19] TC 107-CSP RILEM Recommendation - Measurements of time-dependent strains of concrete. Materials and Structures 31, 8 (1998), 504 - 512.

Appendix A

Development of testing technology

For the conditioning and further testing of the concrete samples at elevated temperatures designing adequate testing technology was required. This appendix describes the equipment developed to accomplish the laboratory investigations. It includes a description of the chambers for conditioning of the samples, the devices used to measure the relative humidity of the concrete pores, and the climate chambers employed for the experiments of creep at elevated temperatures.

A.1 Chambers for conditioning of the samples

The conditioning of the concrete samples at 20 °C was achieved by storing them in four boxes in a climate room at 65 % RH (see Fig. A.1).

Figure A.1: Boxes for storing of the concrete samples at ambient temperature

Two of the boxes were sealed using plexiglass boards on the front and equipped with 6 plastic trays in the bottom. Saturated salt solutions were poured in the trays in order to adjust the relative humidity of the boxes. The selection of the salts and production of the solutions were carried out in accordance with DIN EN ISO 12571 [N11]. An uniform air circulation was ensured by placing radial and axial air blowers in the boxes.

At elevated temperatures the samples were conditioned using the climate chambers shown in Fig. A.2.

Figure A.2: Climate chambers for storing the concrete samples at elevated temperatures

These chambers allowed to increase the temperature at a constant rate and maintain both temperature and relative humidity of the environment according to the desired conditions described in Chapter 3.2. The chambers are able to maintain the temperature in time with an accuracy of ±0.1 °C and the relative humidity remains around ±0.5 % from the adjusted value.

A.2 System to measure the relative humidity of the concrete pores

The relative humidity and temperature of the concrete pores were measured by means of miniaturized temperature and humidity sensors (see Fig.A.3). These sensors, having a diameter of 4 mm, were introduced in concrete samples previously prepared with holes with a diameter of 4.1 mm.

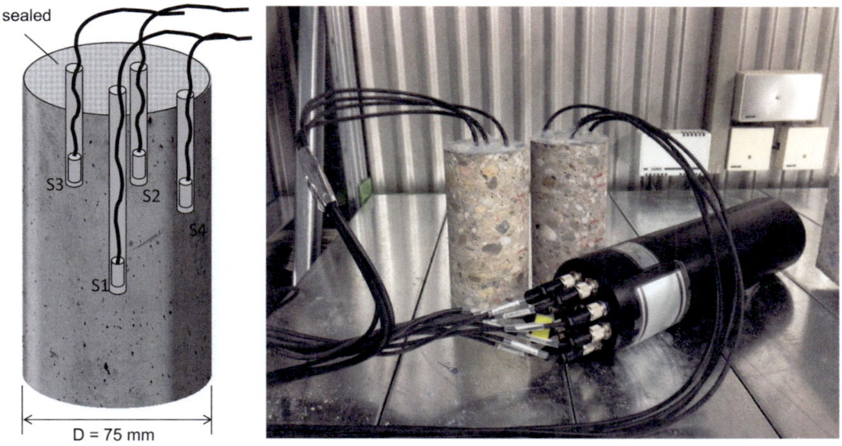

Figure A.3: Sensors placed in the concrete samples to measure the relative humidity and temperature of the concrete pores. Schematic illustration (left) and photo from two samples equipped with sensors connected to the data logger (right)

The measurements were conducted after sealing the top and bottom faces of the samples. The sensors were placed in the samples at the beginning of the conditioning and were left inside during the whole process of conditioning. Several samples were equipped with these sensor in order to measure the different conditioning and testing scenarios. For relative humidities in the range between 5 and 95 %, the sensors guarantee an accuracy of $\pm 2\%$, while for relative humidties close to 0 or 100 %, the accuracy drops to around 5 %. The accuracy of the temperature measures, in the range relevant for the investigation (i.e. from 20 to 80 °C), is 0.5 °C.

A.3 Experimental equipment to measure concrete creep at elevated temperatures

For the experiments of creep at elevated temperatures creep climate chambers were developed. In Fig. A.4 the main components of the chambers are depicted.

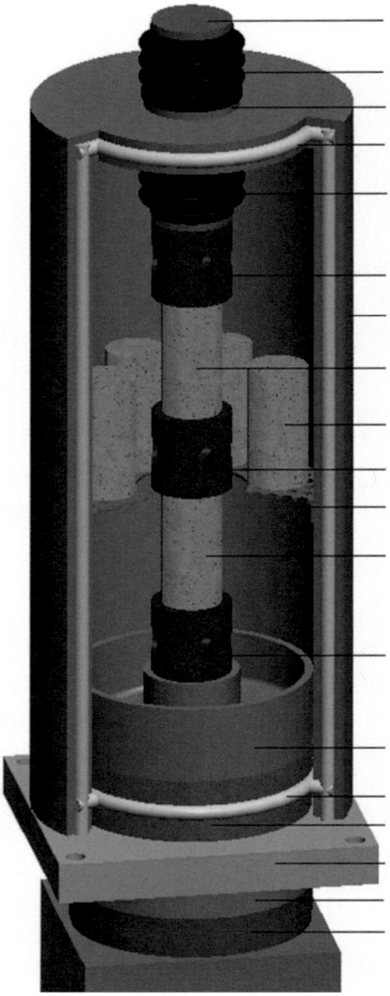

Figure A.4: Chamber to measure concrete creep at elevated temperatures up to 100 °C (door, air blowers and strain measuring devices not included in the figure)

The creep climate chamber consists of a double wall cylinder made out of stainless steel sheets. The temperature in the chambers was controlled by means of electrical heating sleeves placed between the steel sheets. The heating sleeves have an output of 2600 W and can produce a maximum constant temperature of 100 °C in the climatic chamber. The load guide elements direct the load into the samples. In addition, these elements ensure that the temperature of the samples does not vary over the height. Due to the crossed openings, air could circulate and thus the load guide elements were uniformly heated. The control of the relative humidity in the chamber was carried out over saturated salt solutions in the brine container. The deformation of the samples was measured with inductive displacement transducers shown on the right picture of Fig. A.5.

Figure A.5: Preparation of the samples shortly before conducting creep tests at elevated temperatures. Creep climate chambers integrated to the testing machines (left) and detailed view of the samples equipped with inductive displacement transducers (right)

The mechanical pressure force was generated by a creep testing machine with a maximum force of 1000 kN. The climatic chambers were integrated to the creep testing machines as shown by the left picture in Fig. A.5.

A.4 Calculation of the heating rates

In order to prove the functionality of the climate chamber, preliminary tests were carried out. The objective of these tests was to evaluate the development of the temperature and the relative humidity within the chambers during an increment of the temperature. Two chambers were tested at the same time having the same salt solution to control the relative humidity. The air in one of the chamber was left to circulate naturally and the other chamber was equipped with air blowers to force air circulation. In both cases the heating sleeves were set to increase the temperature from 20 to 70 °C with a heating rate of 5 K/h. Each climate chamber was instrumented with five sensors to measure the relative humidity and the temperature of the air. The sensors were located in similar positions in the chambers as shown in Fig. A.6.

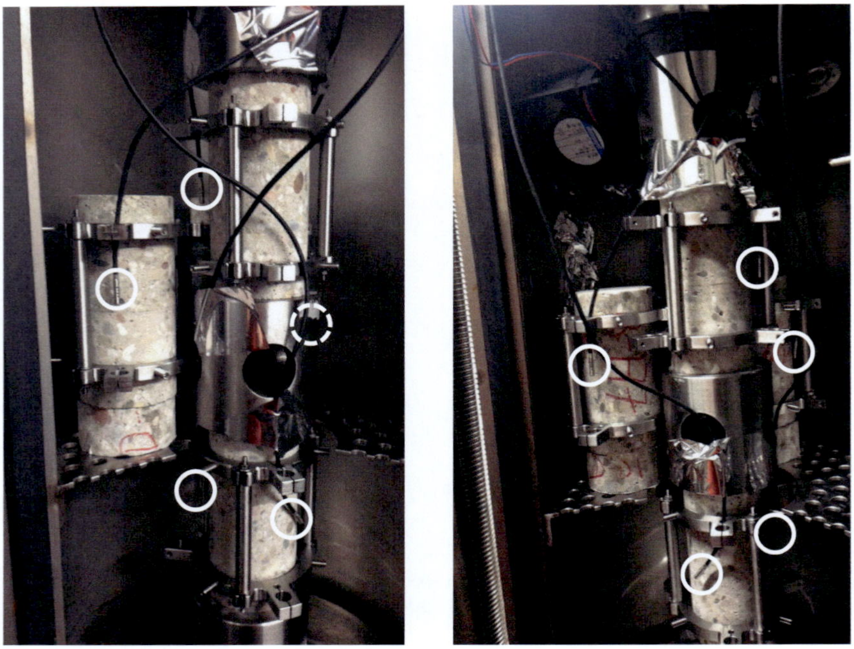

Figure A.6: Locations of the sensors to measure the development of temperature and relative humidity within the creep chambers. Testing the behaviour of chambers without air blowers (left) and with air blowers (right)

The results of the measurements are presented in Figs. A.7 and A.8. Without the use of air blowers the internal air of the chamber did not reach the desired temperature of 70 °C and variations of around ±2 °C in the temperature were measured. With the help of air blowers, the chamber reached the desired temperature and a much better homogeneity in the temperature distribution was achieved (see Fig. A.7).

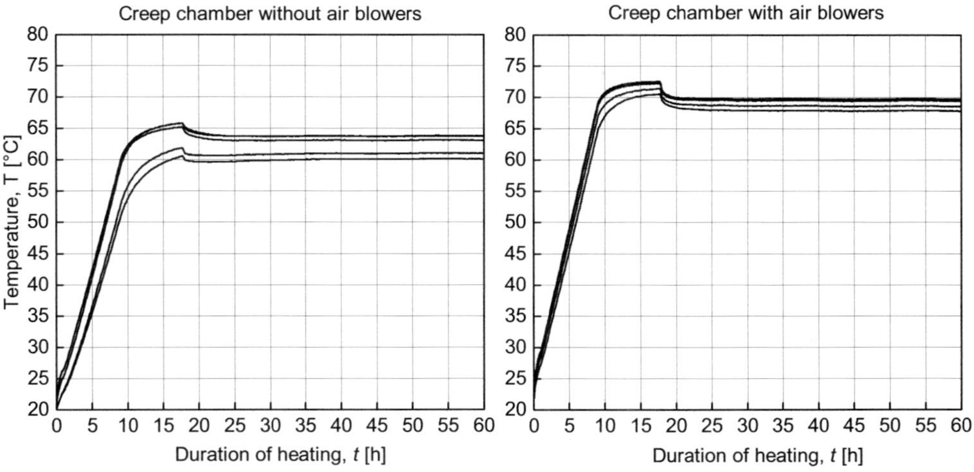

Figure A.7: Measurements of temperature evolution during heating. Comparison between chamber without air blowers (left) and with air blowers (right)

During the heating phase the relative humidity drops down because it takes time for the salt solution to react. This was seen in both cases during the test. However, the use of air blower helped to abate the fall of the relative humidity during the increase of temperature and helped as well to reach a mean relative humidity closer to the desired value of 65 % (see Fig. A.8).

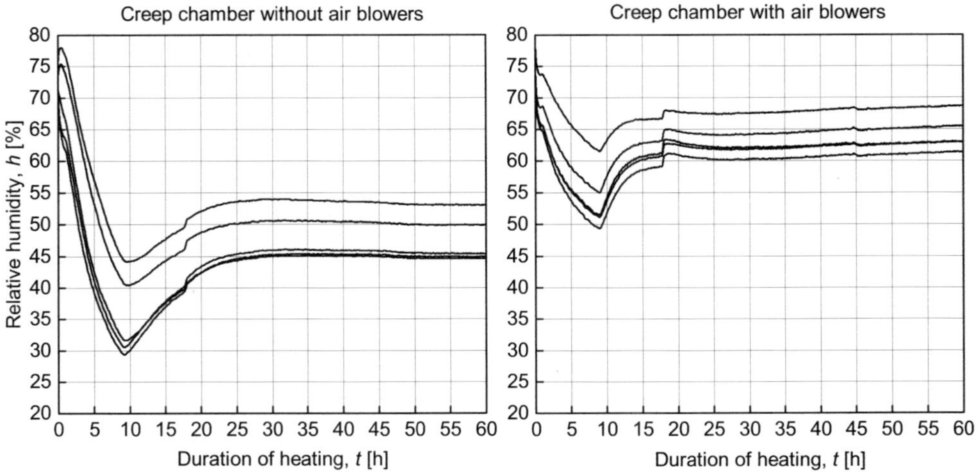

Figure A.8: Measurements of relative humidity evolution during heating. Comparison between chamber without air blowers (left) and with air blowers (right)

Appendix B

Experimental measurements

B.1 Relative humidity and water content in the concrete pores

This section contains the results of the measurements of relative humidity in the concrete pores. The diagrams present the measurements of the development in time of the relative humidity according to the concrete mixture, relative humidity of the surroundings and temperature by drying. In order to calculate the mean values of relative humidity in the samples, the following exponential relation to approximate the humidity profile in function of the radius was used:

$$h = a + (h_\infty - a) \cdot \left(\frac{r}{R}\right)^b \tag{B.1}$$

were a and b are parameters used to calibrate the function, h_∞ is the relative humidity of the surrounding air, r is the position in the sample and R is the radius of the sample. In every point in time, the function B.1 was calibrated using up to four values from the measurements of the sensors at their respective positions and one value corresponding to the relative humidity of the surrounding air at the sample surface. Once the profile of relative humidity distribution from the middle axis to the surface of the sample is approximated, given that the samples have cylindrical geometries, the mean relative humidity of the concrete sample at a given point in time can be calculated by integrating the profile over the sample area and dividing it by the total area as follows.

$$h_{\mathrm{mean}} = \frac{\int h \cdot dA}{A} = \frac{\int_{r=0}^{r=R} h \cdot 2\pi r \cdot dr}{\pi \cdot R^2} \approx a + 2 \cdot \frac{h_\infty - a}{b+2} \tag{B.2}$$

For the samples used to determine the tensile strength, the presence of the notch can be taken into account by reducing the sample radius to $R-t$. The mean relative humidity of the samples type 2 is then calculated as

$$h_{\text{mean}} = \frac{\int h \cdot dA}{A} = \frac{\int\limits_{r=0}^{r=R-t} h \cdot 2\pi r \cdot dr}{\pi \cdot (R-t)^2} \approx a + 2 \cdot \frac{h_\infty - a}{b+2} \cdot \frac{(R-t)^b}{R^b} \tag{B.3}$$

where t is the notch depth.

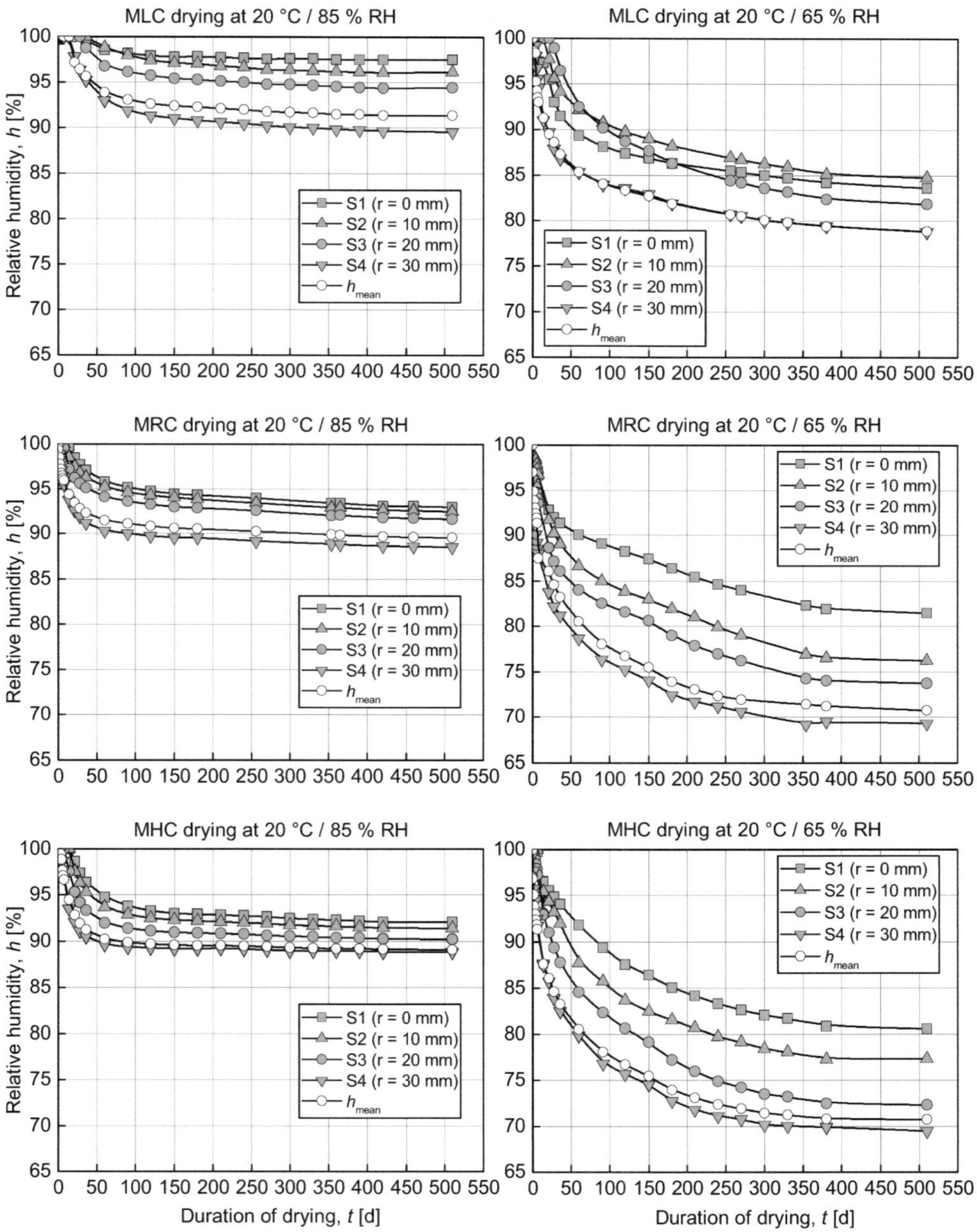

Figure B.1: Measurements of relative humidity in the concrete MLC, MRC, and MHC during drying at 20 °C / 85 % RH (left) and 20 °C / 65 % RH (right)

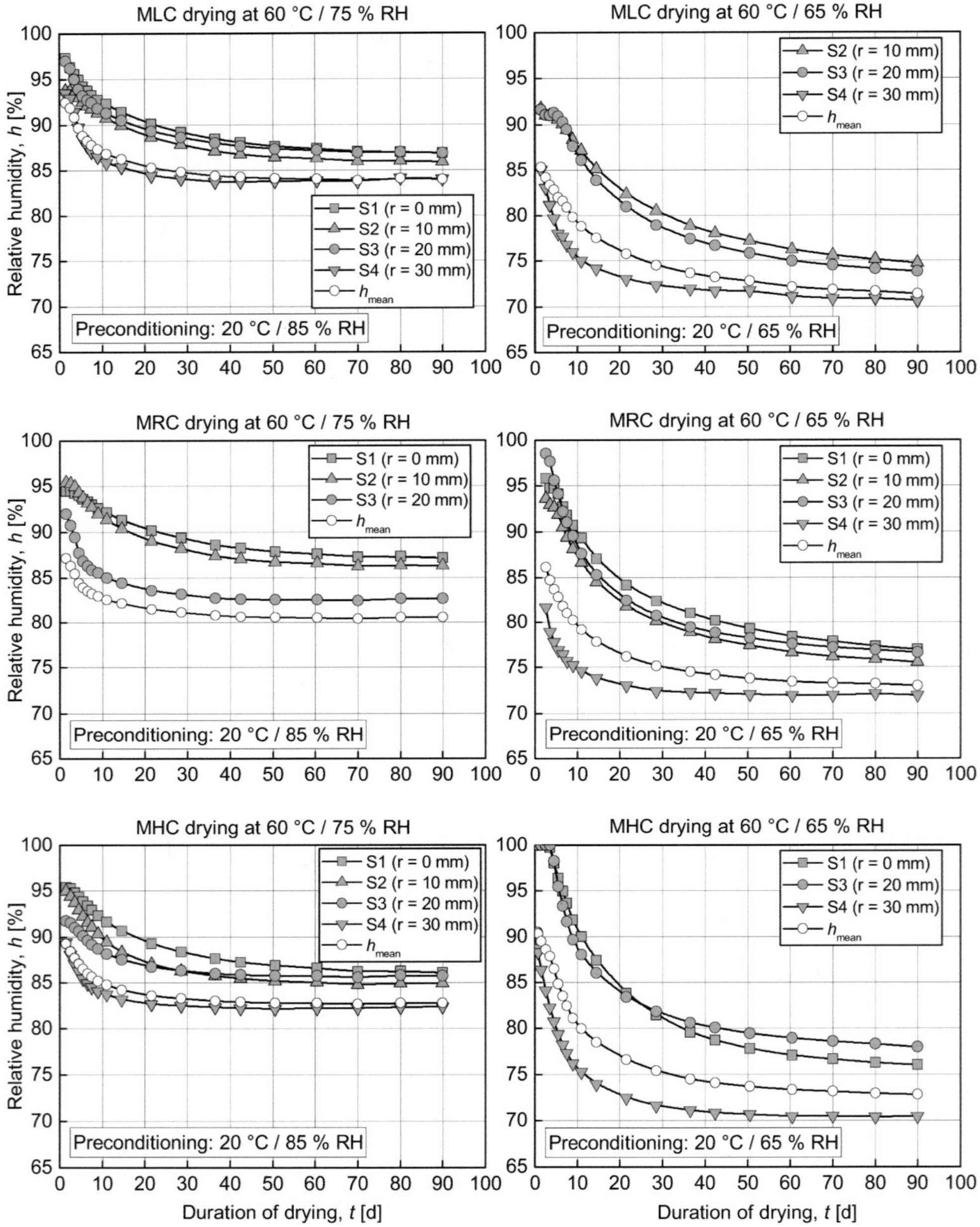

Figure B.2: Measurements of relative humidity in the concrete MLC, MRC, and MHC during drying at 60 °C / 75 % RH (left) and 60 °C / 65 % RH (right)

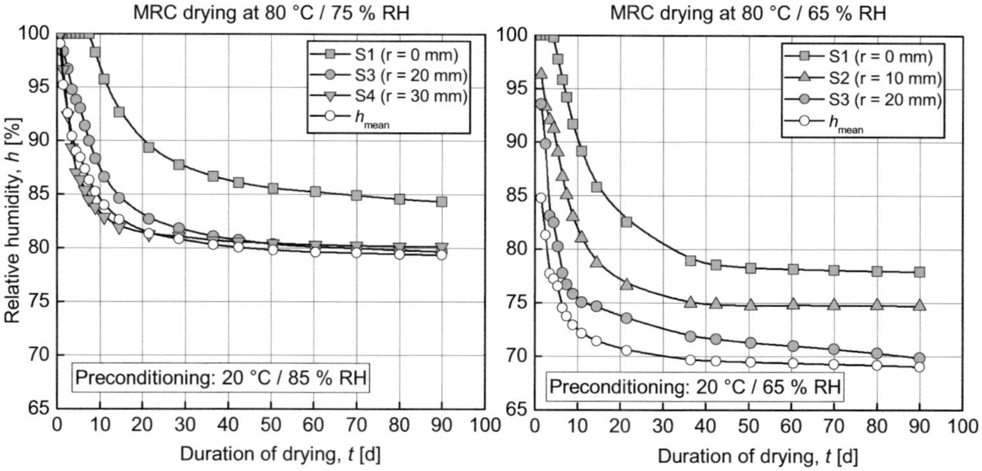

Figure B.3: Measurements of relative humidity in the concrete MRC during drying at 80 °C / 75 % RH (left) and 80 °C / 65 % RH (right)

During the measurements of relative humidity conducted at elevated temperatures some of the humidity sensors failed and therefore in some of the diagrams presented in Figs. B.2 and B.3 the measured results from only three sensors are given. In these cases the calculation of the mean values followed the same procedure describe in Section B.1, with the difference that only four values (three measured by the sensors and one corresponding to the relative humidity of the surrounding) were used to calculate the parameters a and b from Eq. B.1.

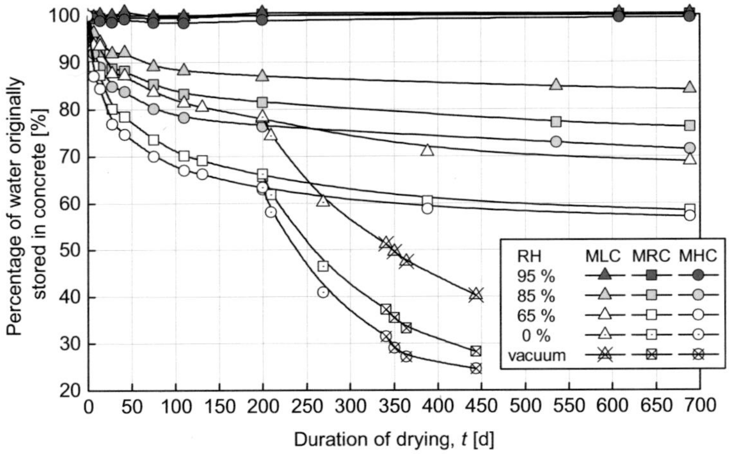

Figure B.4: Time development of water stored in concretes MLC, MRC and MHC in relation to the water stored before the beginning of drying at 20 °C and different ambient relative humidities

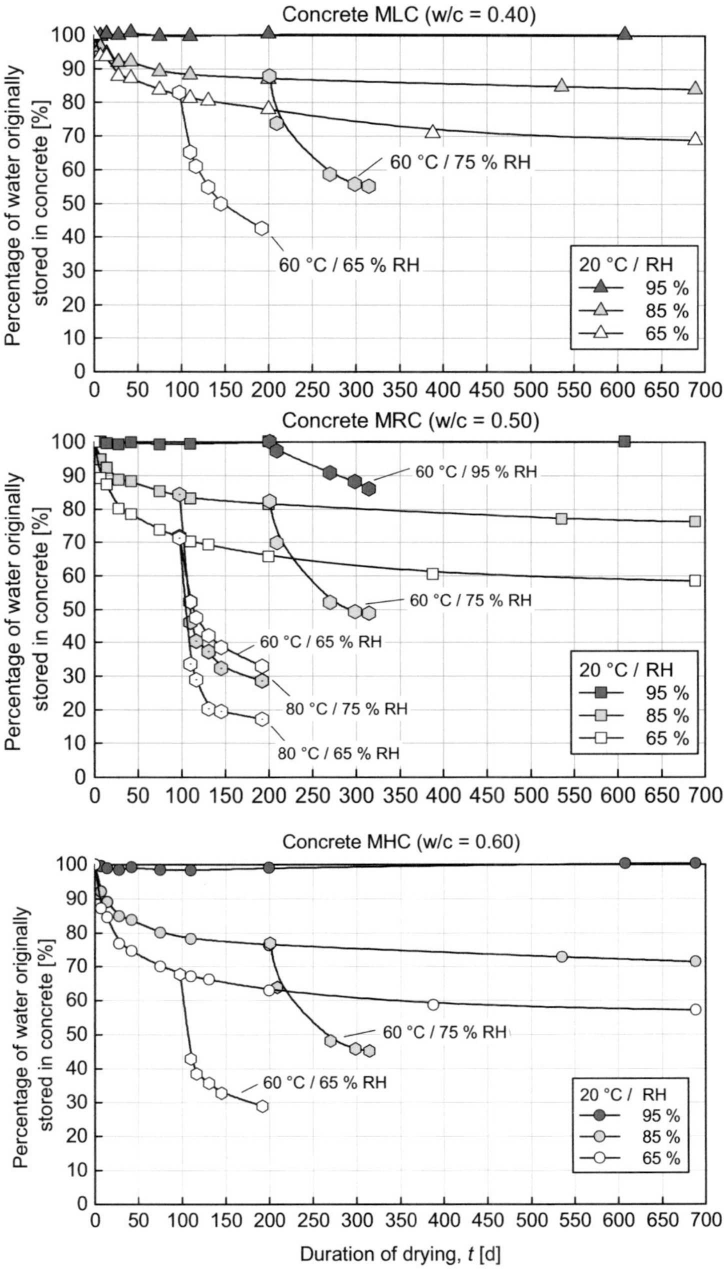

Figure B.5: Time development of water stored in the concretes MLC, MRC and MHC in relation to the water stored before the beginning of drying at 20, 60 and 80 °C and different ambient relative humidities

B.2 Mercury intrusion and gas permeability

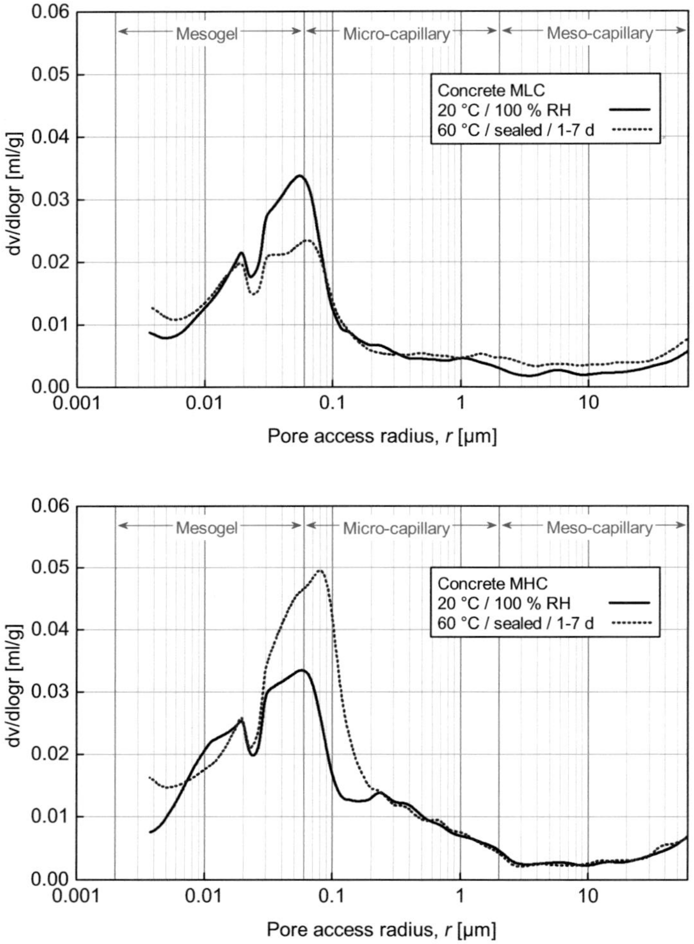

Figure B.6: Pore size distribution of concretes MLC (top) and MHC (bottom) dried at 20 °C / 85 % RH and then subject to 60 °C

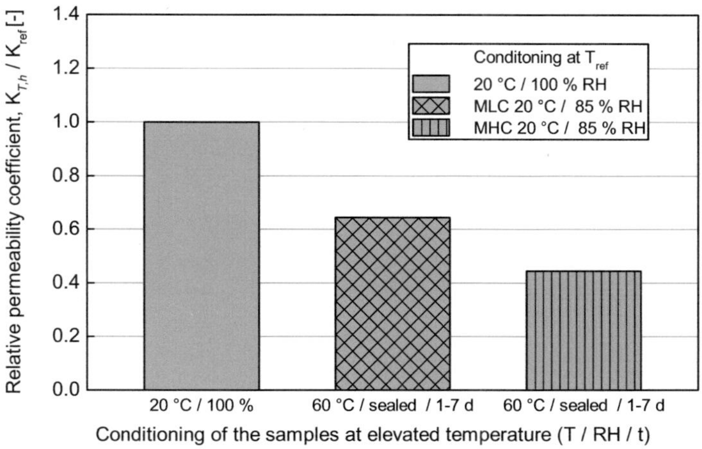

Figure B.7: Permeability coefficients of concretes MLC and MHC previously stored at 20 °C / 85 % RH and then subject to 60 °C in relation to the permeability coefficients measured at reference conditions

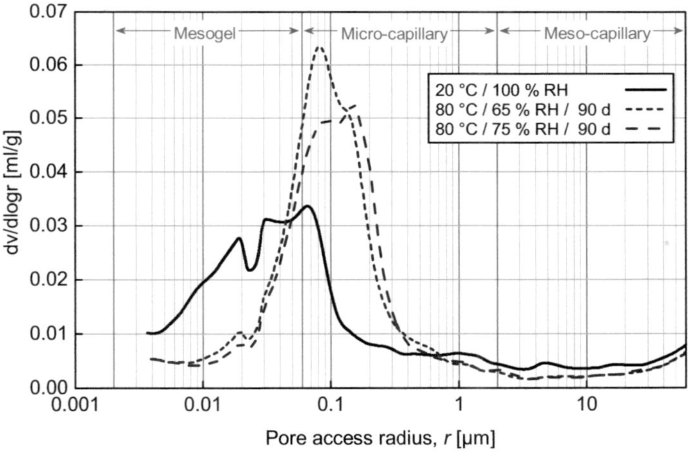

Figure B.8: Pore size distribution of concrete MRC dried at 20 °C / 65 % RH and then subject to 80 °C / 65 % RH and dried at 20 °C / 85 % RH and then subject to 80 °C / 75 % RH

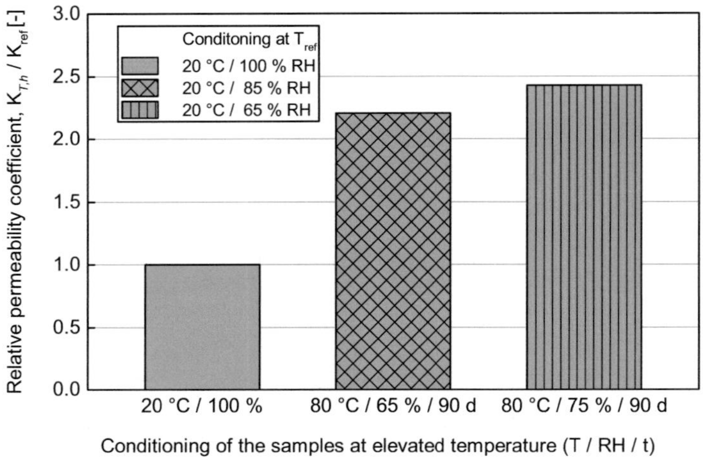

Figure B.9: Permeability coefficients of concrete MRC subject 80 °C and different humidity conditions in relation to the permeability coefficient measured at reference conditions

B.3 Concrete strength and stiffness

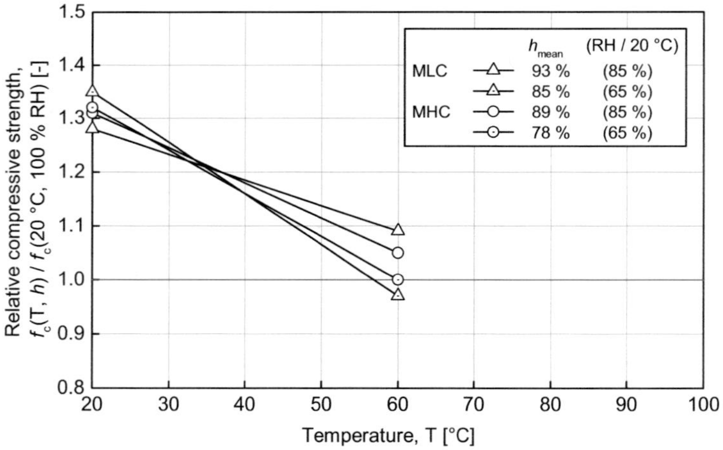

Figure B.10: Relative compressive strength of the concrete mixtures MLC and MHC heated at different moisture contents

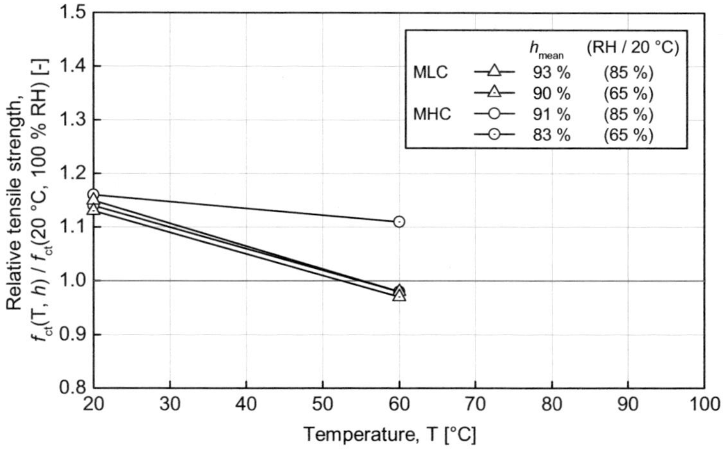

Figure B.11: Relative tensile strength of the concrete mixtures MLC and MHC heated at different moisture contents

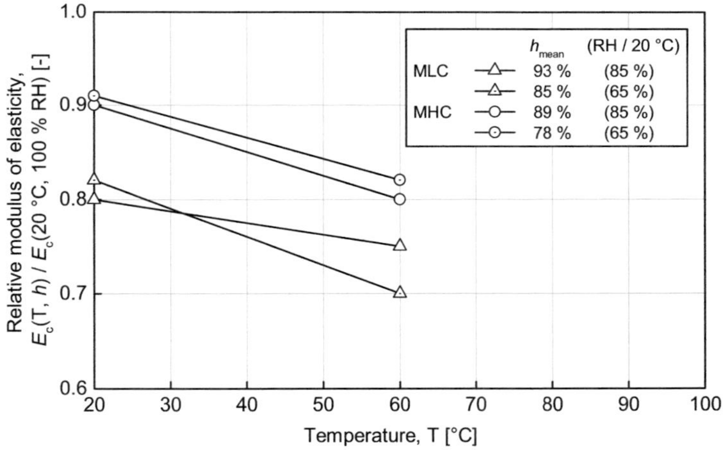

Figure B.12: Relative modulus of elasticity of the concrete mixtures MLC and MHC heated at different moisture contents

B.4 Concrete creep and shrinkage

This section contains the results of the conducted experiments of creep and shrinkage. The main features of the experiments are summarized in the following table.

Table B.1: Load applied on the creep samples

Concrete Mixture	Conditioning at 20 °C	Conditioning during loading		Applied load	Elastic deformation
	RH	RH	Temperature	σ_c	ε_{el}
	[%]	[%]	[°C]	[N/mm²]	[μm/m]
MRC	65	65	20	19.2	528
			40	18.6	455
			70	15.2	399
	85	65	20	19.2	507
			40	19.9	502
			70	17.2	467
	95	65	20	19.1	503
			70	17.2	517
MLC	65	65	20	26.4	670
			70	19.3	520
	85	65	20	26.4	673
			70	19.3	515
MHC	65	65	20	18.0	519
			70	13.2	417
	85	65	20	18.0	509
			70	13.2	413

B.4.1 Measurements of relative humidity during the creep tests conducted at reference temperature

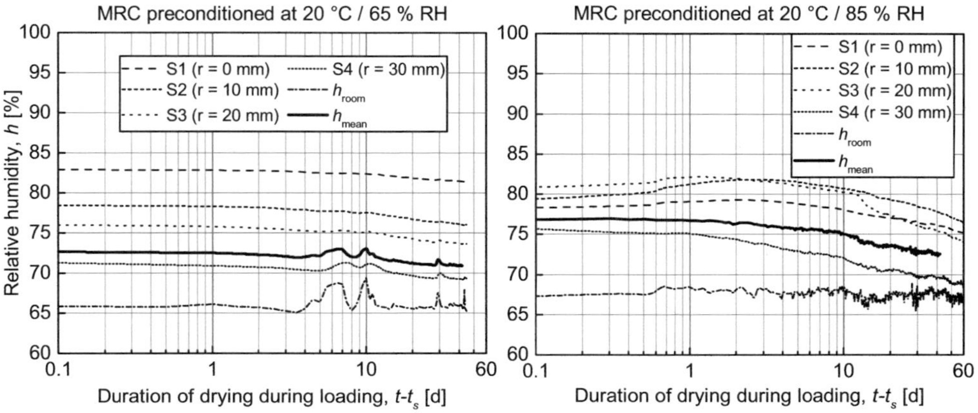

Figure B.13: Pore relative humidity of the concrete MRC during the creep test conducted at 20 °C and 65 % RH for the preconditions 20 °C / 65 % RH (left) and 20 °C / 85 % RH (right)

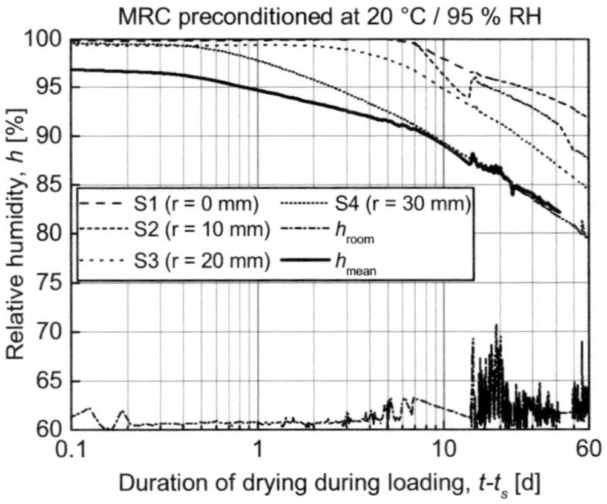

Figure B.14: Pore relative humidity of the concrete MRC during the creep test conducted at 20 °C and 65 % RH for the precondition 20 °C / 95 % RH

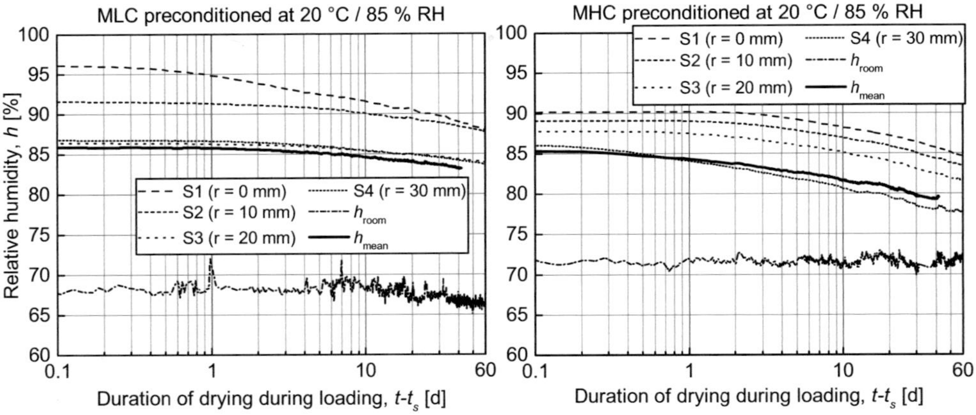

Figure B.15: Pore relative humidity of the concretes MLC (left) and MHC (right) during the creep test conducted at 20 °C and 65 % RH for the precondition 20 °C / 85 % RH

B.4.2 Measurements of relative humidity during the creep tests conducted at elevated temperatures

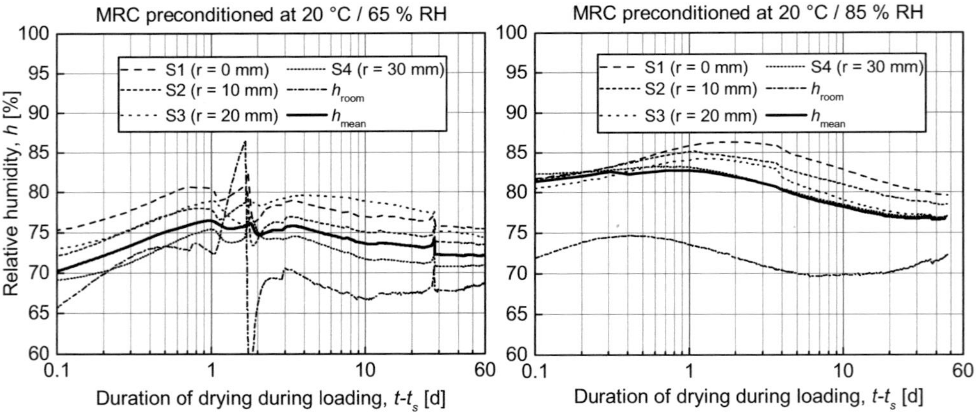

Figure B.16: Pore relative humidity of the concrete MRC during the creep test conducted at 40 °C and 65 % RH for the preconditions 20 °C / 65 % RH (left) and 20 °C / 85 % RH (right)

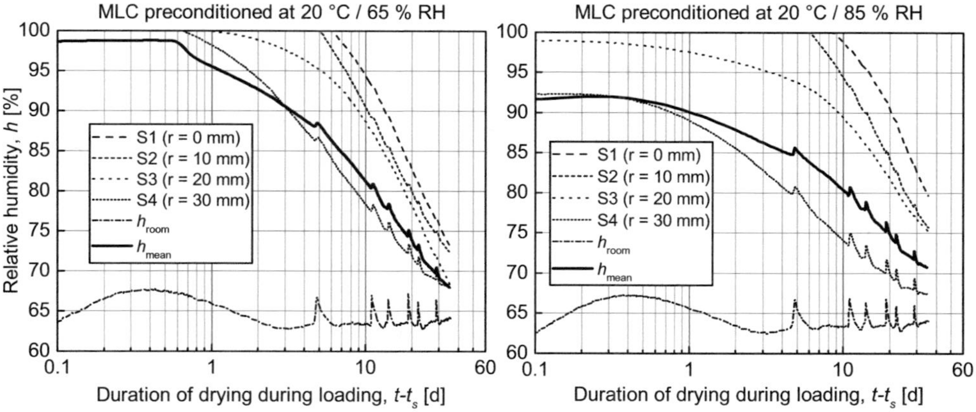

Figure B.17: Pore relative humidity of the concrete MLC during the creep test conducted at 70 °C and 65 % RH for the preconditions 20 °C / 65 % RH (left) and 20 °C / 85 % RH (right)

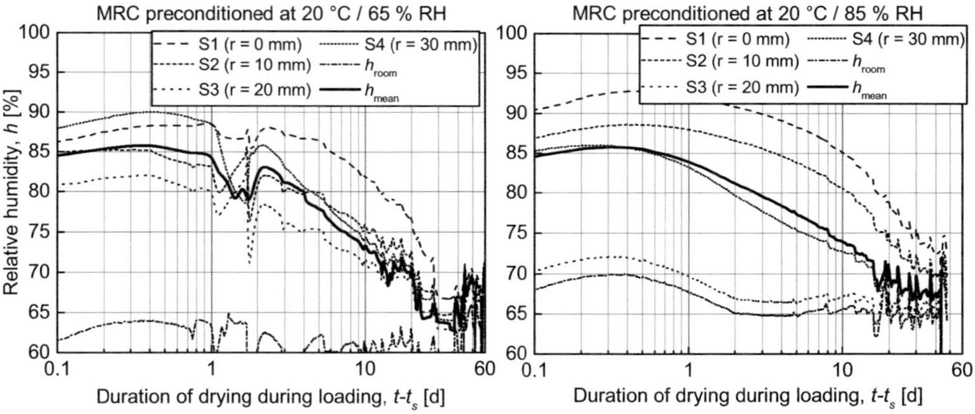

Figure B.18: Pore relative humidity of the concrete MRC during the creep test conducted at 70 °C and 65 % RH for the preconditions 20 °C / 65 % RH (left) and 20 °C / 85 % RH (right)

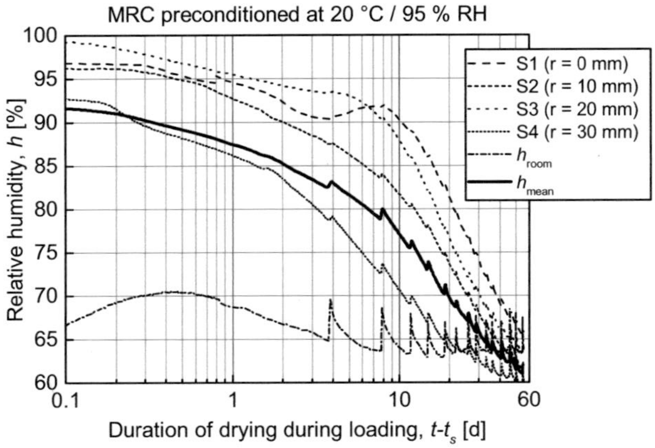

Figure B.19: Pore relative humidity of the concrete MRC during the creep test conducted at 70 °C and 65 % RH for the precondition 20 °C / 95 % RH

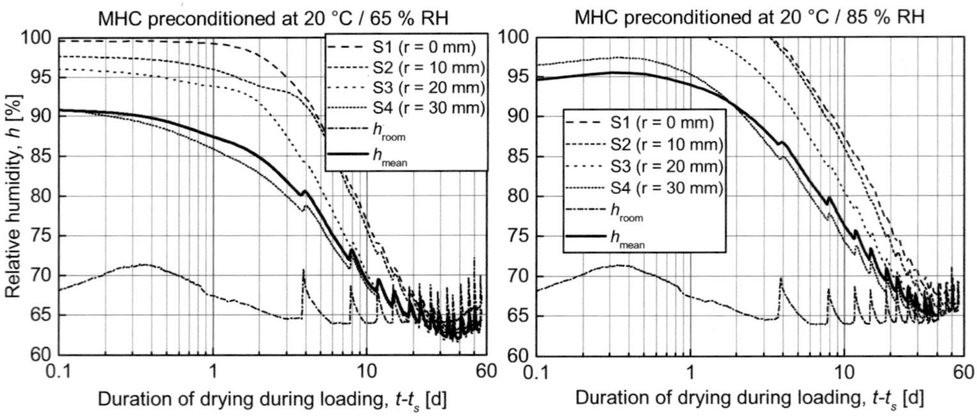

Figure B.20: Pore relative humidity of the concrete MHC during the creep test conducted at 70 °C and 65 % RH for the preconditions 20 °C / 65 % RH (left) and 20 °C / 85 % RH (right)

B.4.3 Total deformations of the samples tested at reference temperature

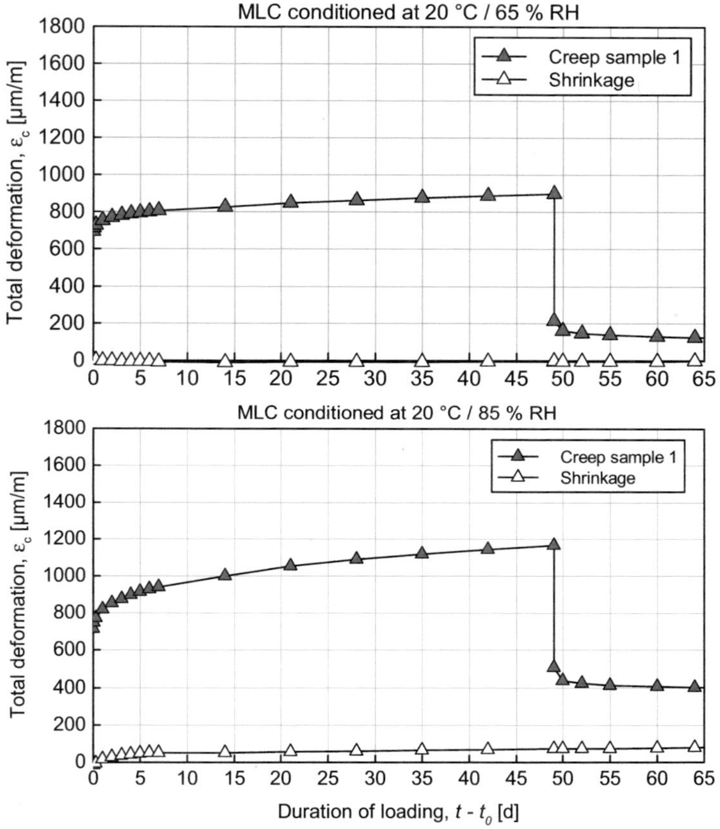

Figure B.21: Time development of the total deformation of concrete MLC tested at 20 °C / 65 % RH according to the conditioning prior to test: 20 °C / 65 % RH (top) and 85 % RH (bottom)

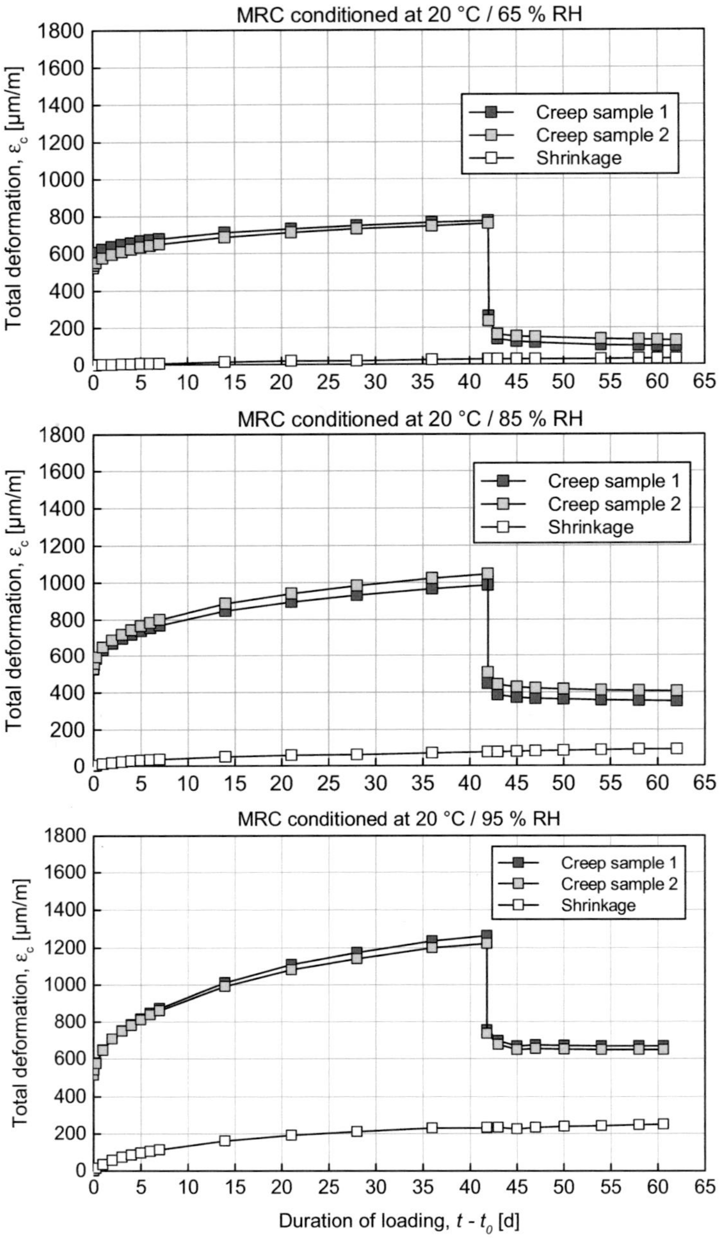

Figure B.22: Time development of the total deformation of concrete MRC tested at 20 °C / 65 % RH according to the conditioning prior to test: 20 °C / 65 % RH (top), 85 % RH (middle), and 95 % RH (bottom)

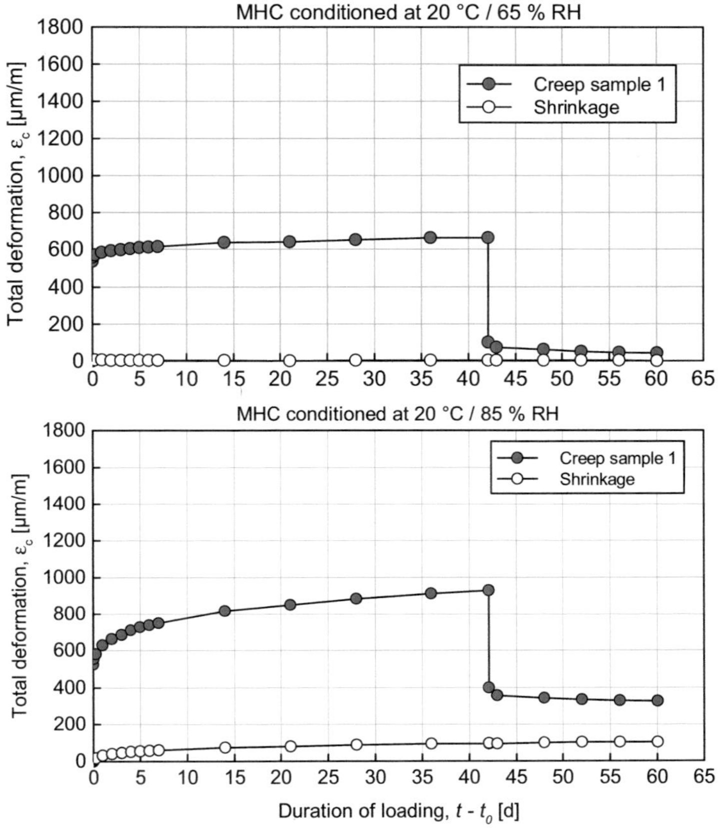

Figure B.23: Time development of the total deformation of concrete MHC tested at 20 °C / 65 % RH according to the conditioning prior to test: 20 °C / 65 % RH (top) and 85 % RH (bottom)

B.4.4 Total deformations of the samples tested at elevated temperatures

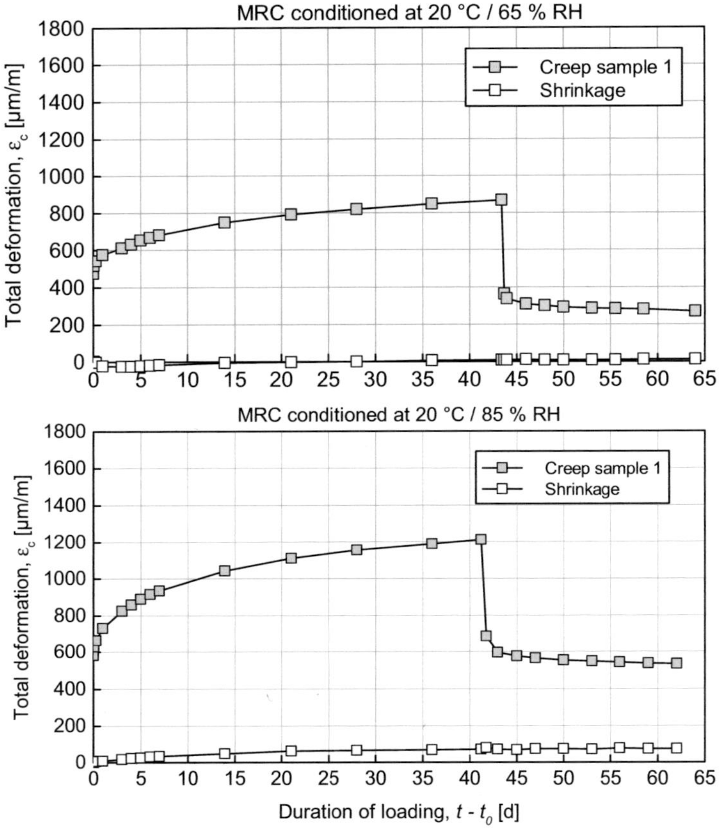

Figure B.24: Time development of the total deformation of concrete MRC tested at 40 °C / 65 % RH according to the conditioning prior to test: 20 °C / 65 % RH (top) and 85 % RH (bottom)

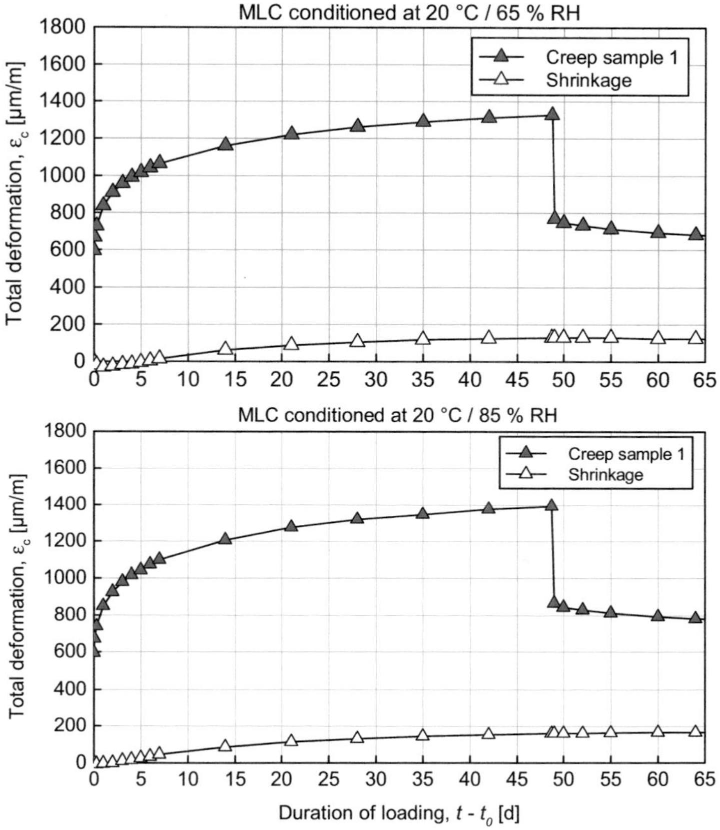

Figure B.25: Time development of the total deformation of concrete MLC tested at 70 °C / 65 % RH according to the conditioning prior to test: 20 °C / 65 % RH (top) and 85 % RH (bottom)

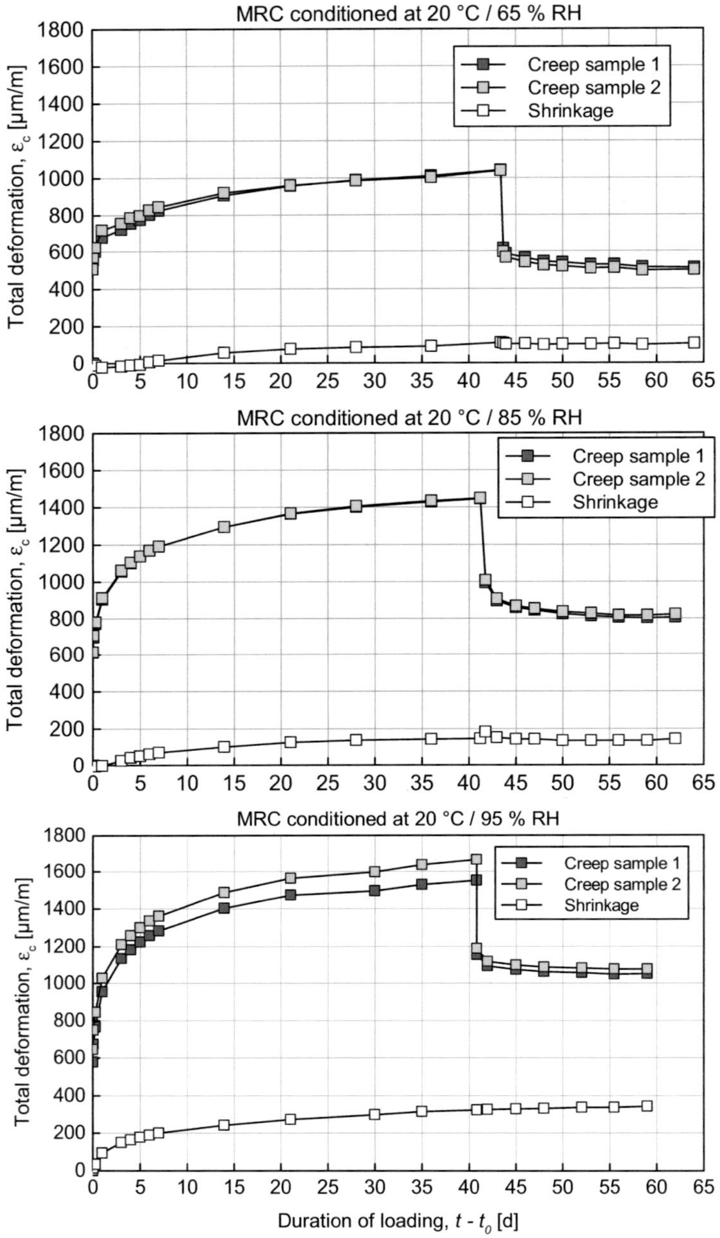

Figure B.26: Time development of the total deformation of concrete MRC tested at 70 °C / 65 % RH according to the conditioning prior to test: 20 °C / 65 % RH (top), 85 % RH (middle), and 95 % RH (bottom)

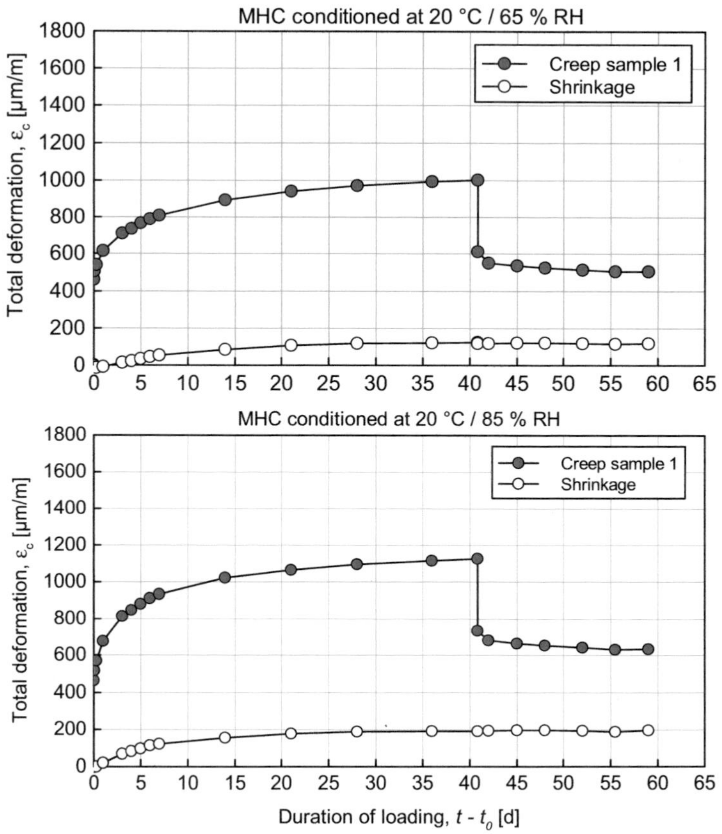

Figure B.27: Time development of the total deformation of concrete MHC tested at 70 °C / 65 % RH according to the conditioning prior to test: 20 °C / 65 % RH (top) and 85 % RH (bottom)

B.4.5 Shrinkage deformations of the samples tested at reference temperature

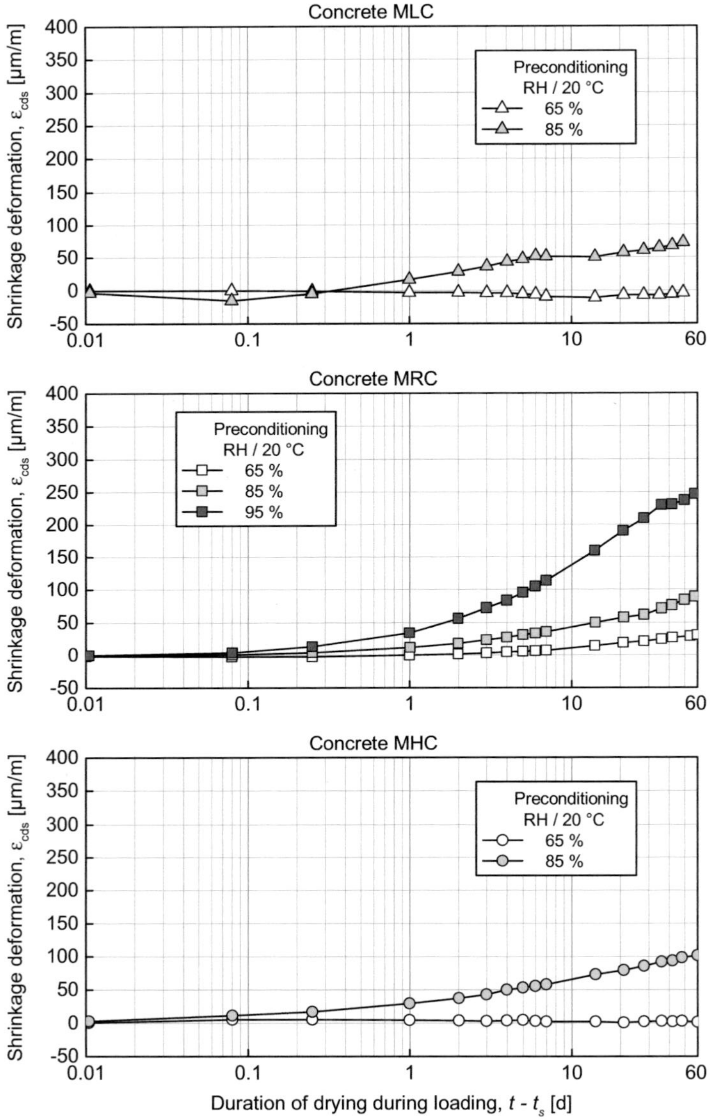

Figure B.28: Time development of the shrinkage deformation of the concretes MLC (top), MRC (middle) and MHC (bottom) tested at 20 °C / 65 % RH according to the conditioning prior to test

B.4.6 Shrinkage deformations of the samples tested at elevated temperatures

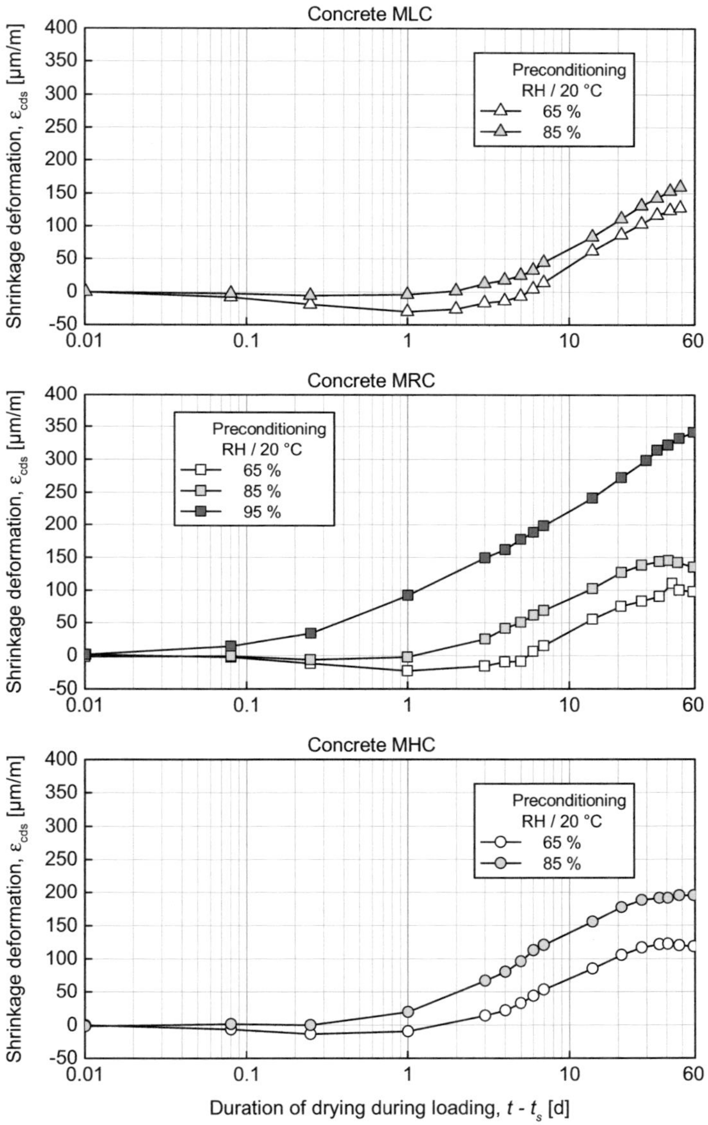

Figure B.29: Time development of the shrinkage deformation of the concretes MLC (top), MRC (middle) and MHC (bottom) tested at 70 °C / 65 % RH according to the conditioning prior to test

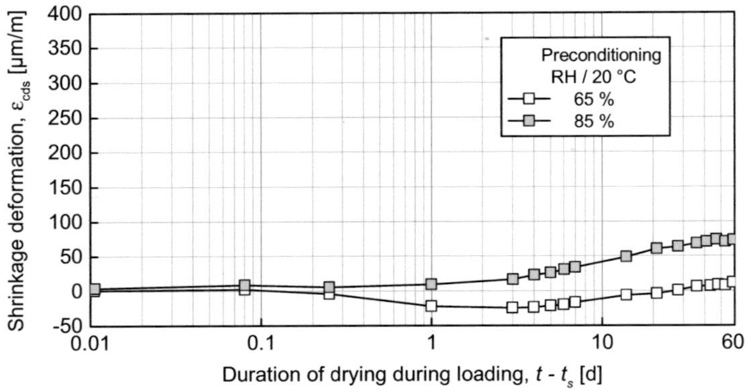

Figure B.30: Time development of the shrinkage deformation of the concrete MRC tested at 40 °C / 65 % RH according to the conditioning prior to test

B.4.7 Specific creep of the concrete samples from tests conducted at reference temperature

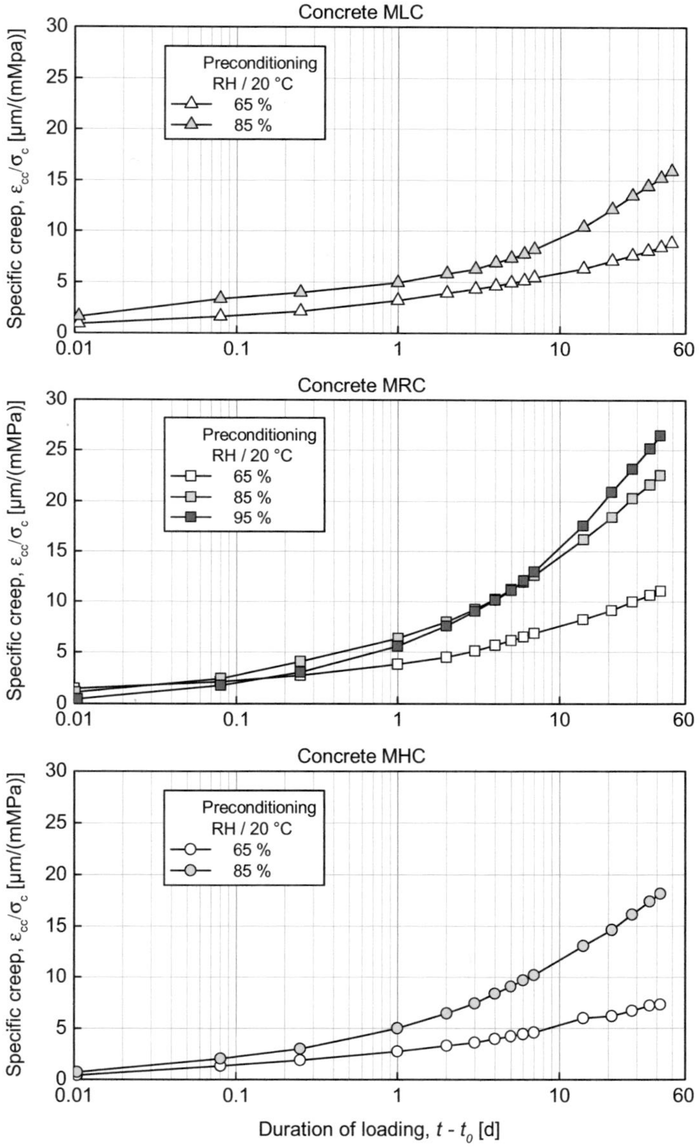

Figure B.31: Time development of the specific creep of the concretes MLC (top), MRC (middle) and MHC (bottom) tested at 20 °C / 65 % RH according to the conditioning prior to test

B.4.8 Specific creep of the concrete samples from tests conducted at elevated temperatures

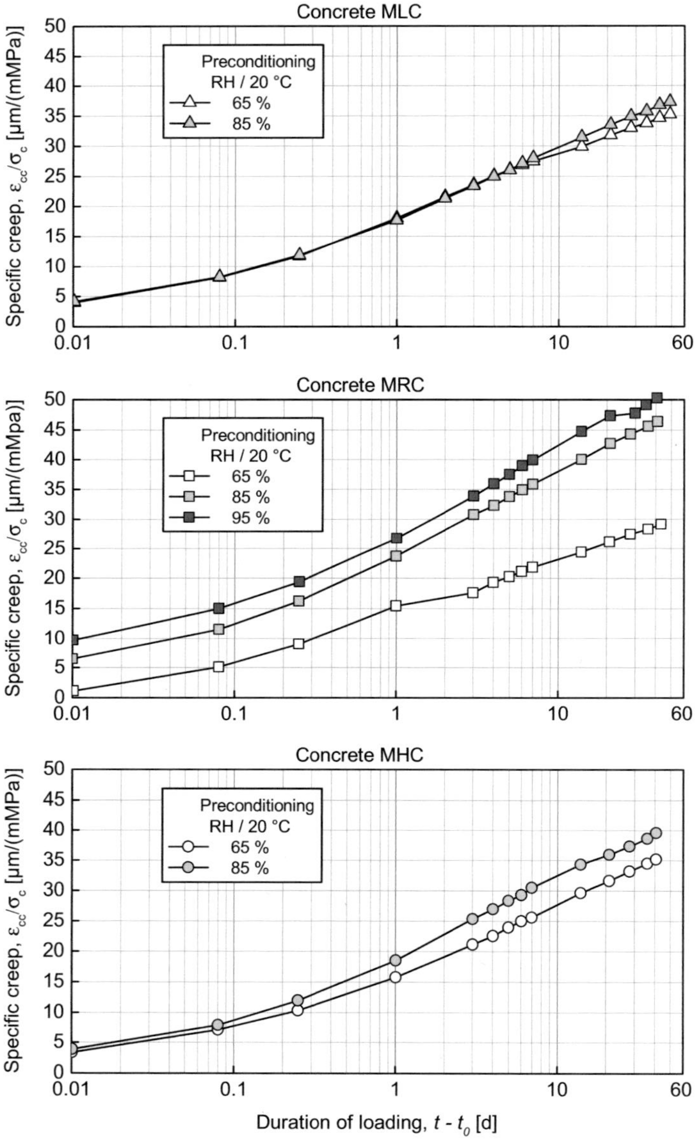

Figure B.32: Time development of the specific creep of the concretes MLC (top), MRC (middle) and MHC (bottom) tested at 70 °C / 65 % RH according to the conditioning prior to test

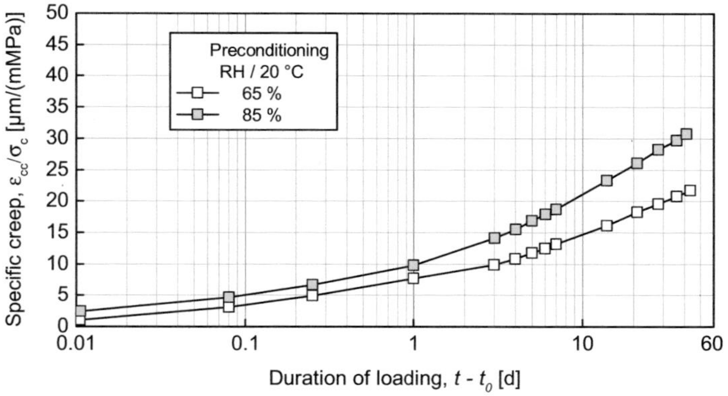

Figure B.33: Time development of the specific creep of the concrete MRC tested at 40 °C / 65 % RH according to the conditioning prior to test

Appendix C

Additional calculations with the developed basic equations

C.1 Concrete strength and stiffness

The following diagrams are graphical representations of the equations to calculate the influence of thermal incompatibilities and drying on the compressive strength, tensile strength and modulus of elasticity of concrete according to the formulations presented in Chapter 5.1.

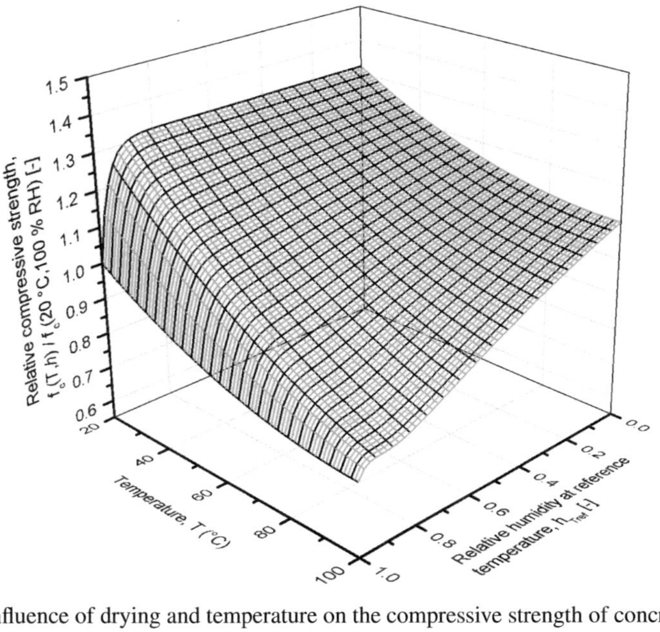

Figure C.1: Influence of drying and temperature on the compressive strength of concrete according to the proposed formulations

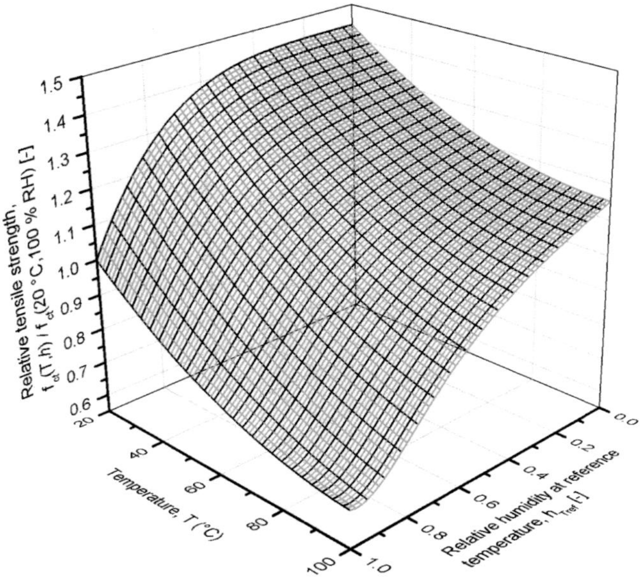

Figure C.2: Influence of drying and temperature on the tensile strength of concrete according to the proposed formulations

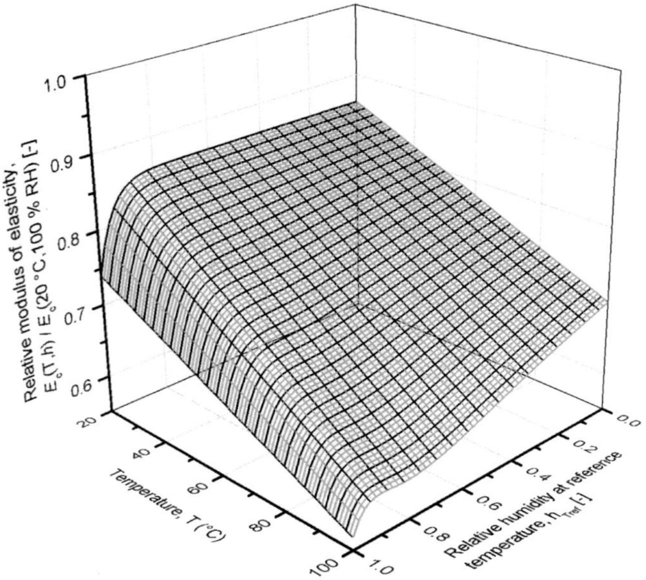

Figure C.3: Influence of drying and temperature on the modulus of elasticity of concrete according to the proposed formulations

C.2 Concrete shrinkage

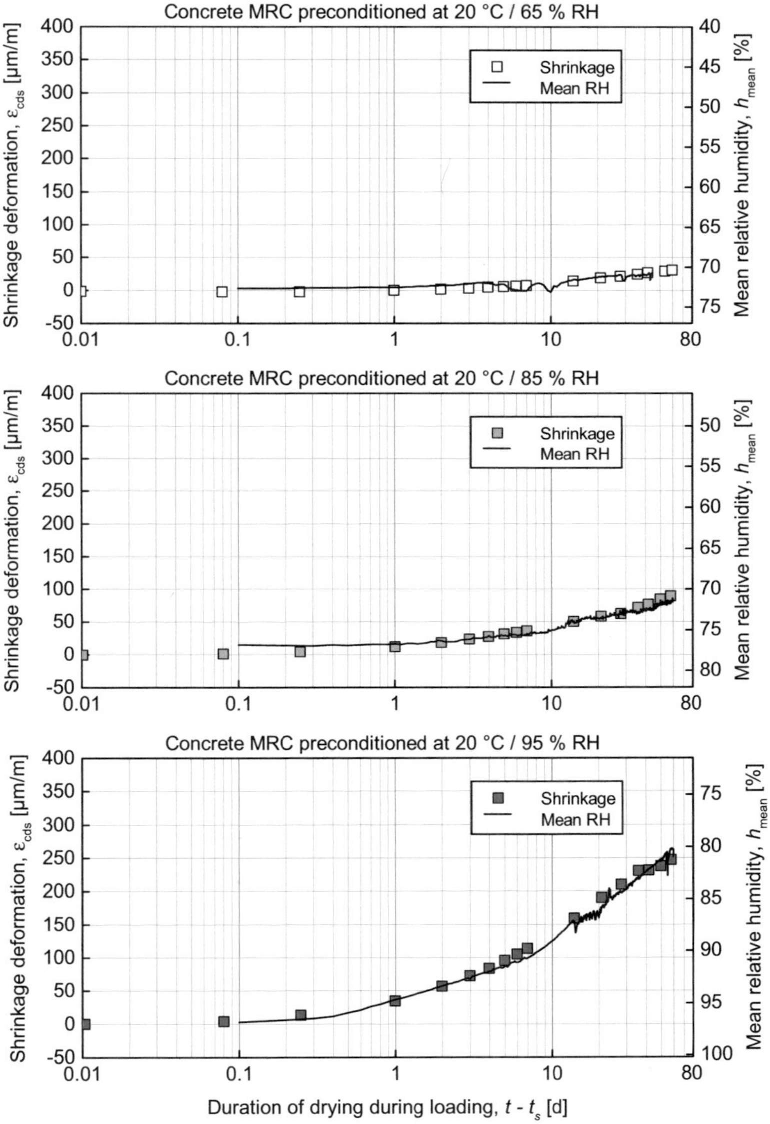

Figure C.4: Shrinkage deformation and mean relative humidity of the samples from the concrete MRC previously conditioned at 20 °C and 65 %, 85 % and 95 % RH during the creep tests conducted at 20 °C and 65 % RH

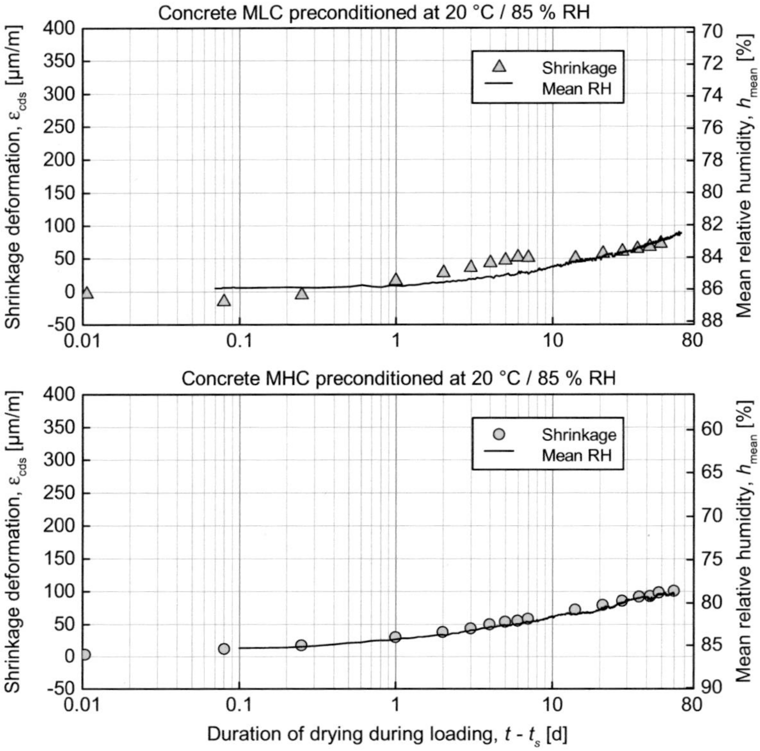

Figure C.5: Shrinkage deformation and mean relative humidity of the samples from the concrete MLC and MHC previously conditioned at 20 °C and 85 % RH during the creep tests conducted at 20 °C and 65 % RH

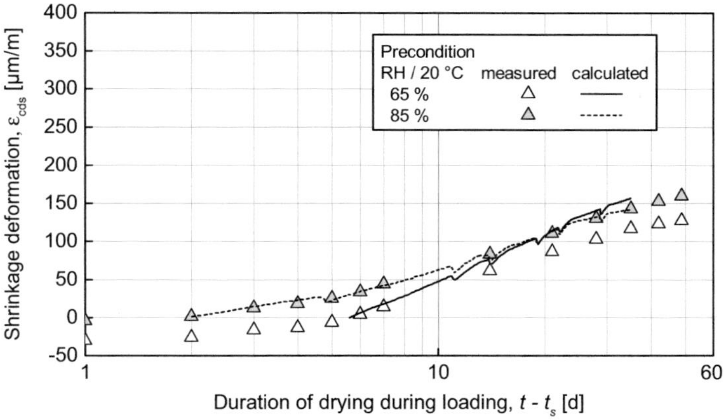

Figure C.6: Shrinkage calculated with the mean relative humidity measured in the concrete pores in comparison with the measured values for the concrete MLC during the creep tests conducted at 70 °C and 65 % RH

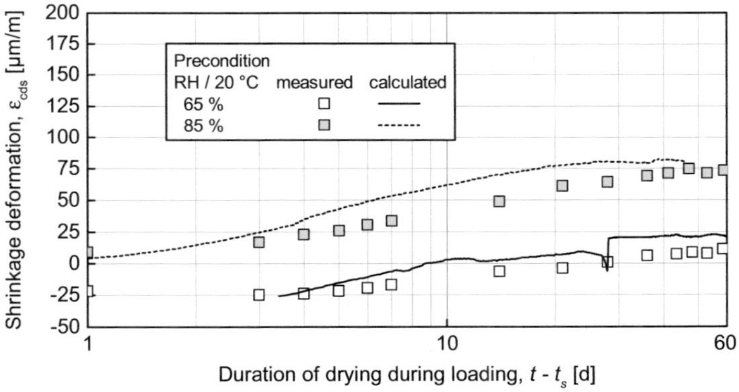

Figure C.7: Shrinkage calculated with the mean relative humidity measured in the concrete pores in comparison with the measured values for the concretes MRC during the creep tests conducted at 40 °C and 65 % RH

C.3 Concrete creep

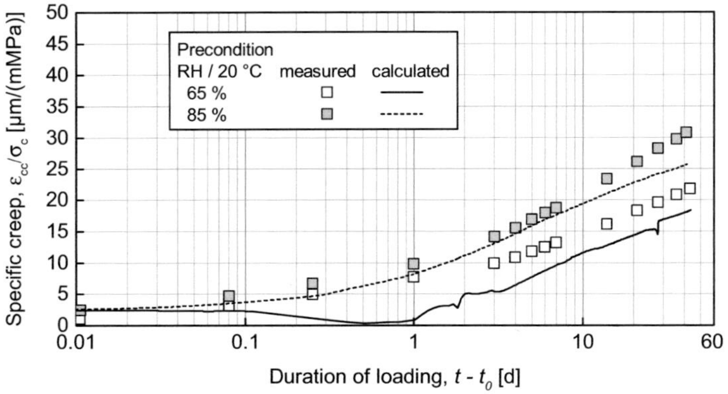

Figure C.8: Specific creep calculated with the mean relative humidity measured in the concrete pores in comparison with the measured values for the concretes MRC during the creep tests conducted at 40 °C and 65 % RH

Appendix D

Additional calculations with the new overall material model

D.1 Moisture transport at constant conditions of temperature and humidity

The following diagrams present comparison between the time-development of the mean relative humidity in the concrete pores calculated with the moisture transport model and according to the measurements for the concrete MRC drying at 20, 60, and 80 °C.

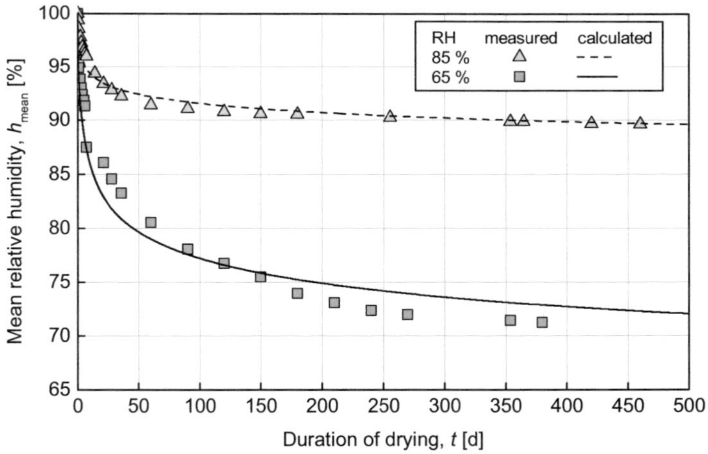

Figure D.1: Comparisons between the measured values of concrete MRC drying at 20 °C and 85 % as well as 65 % RH and the calculated values using the model of moisture transport

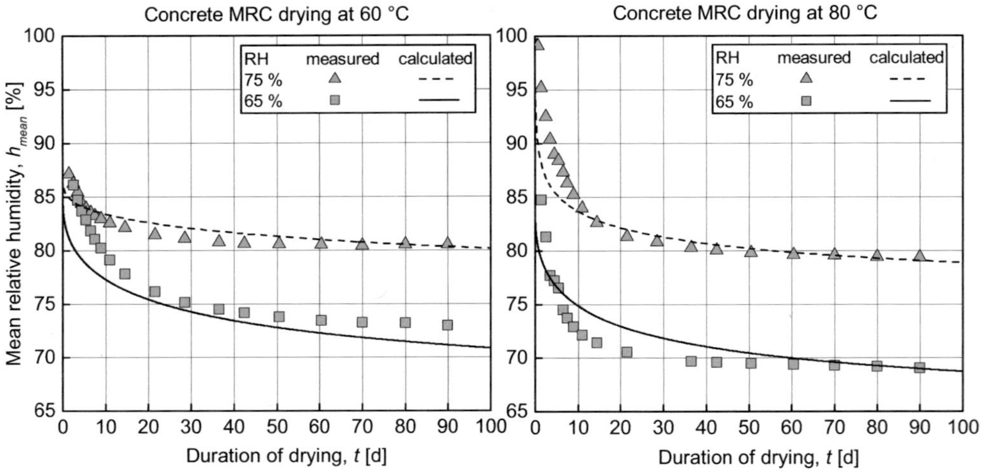

Figure D.2: Comparisons between the calculated values using the model of moisture transport and the measured values of mean relative humidity from the concrete MRC drying at 60 °C (left) and 80 °C (right) at relative humidities of 75 and 65 %

D.2 Concrete strength and stiffness

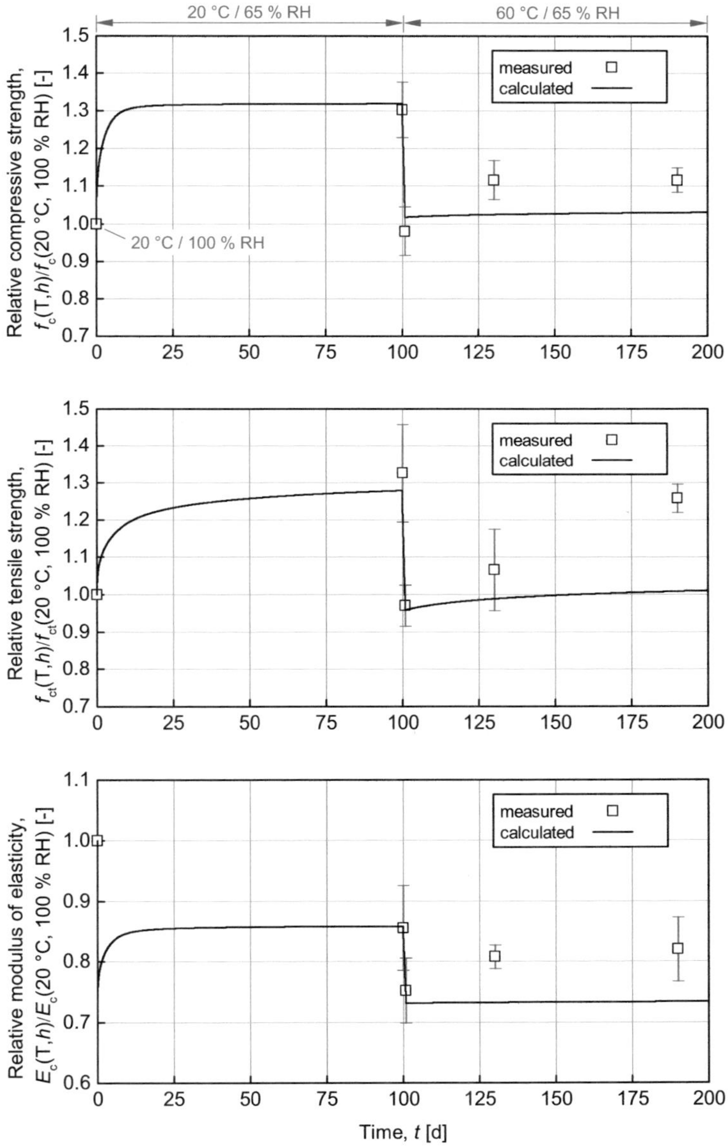

Figure D.3: Comparison between calculations with the material model and measurements of compressive strength (top), tensile strength (middle) and modulus of elasticity (bottom) for the concrete MRC after drying during 100 d at 20 °C / 65 % RH and then heated during 100 d at 60 °C / 65% RH

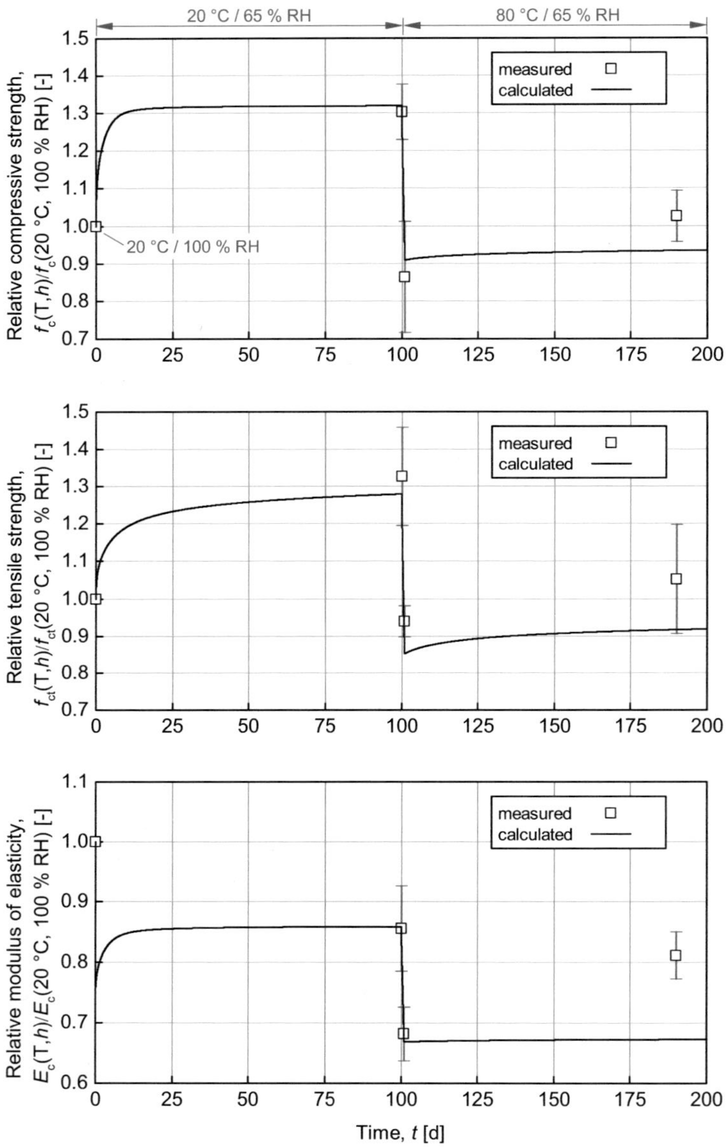

Figure D.4: Comparison between calculations with the material model and measurements of compressive strength (top), tensile strength (middle) and modulus of elasticity (bottom) for the concrete MRC after drying during 100 d at 20 °C / 65 % RH and then heated during 100 d at 80 °C / 65% RH

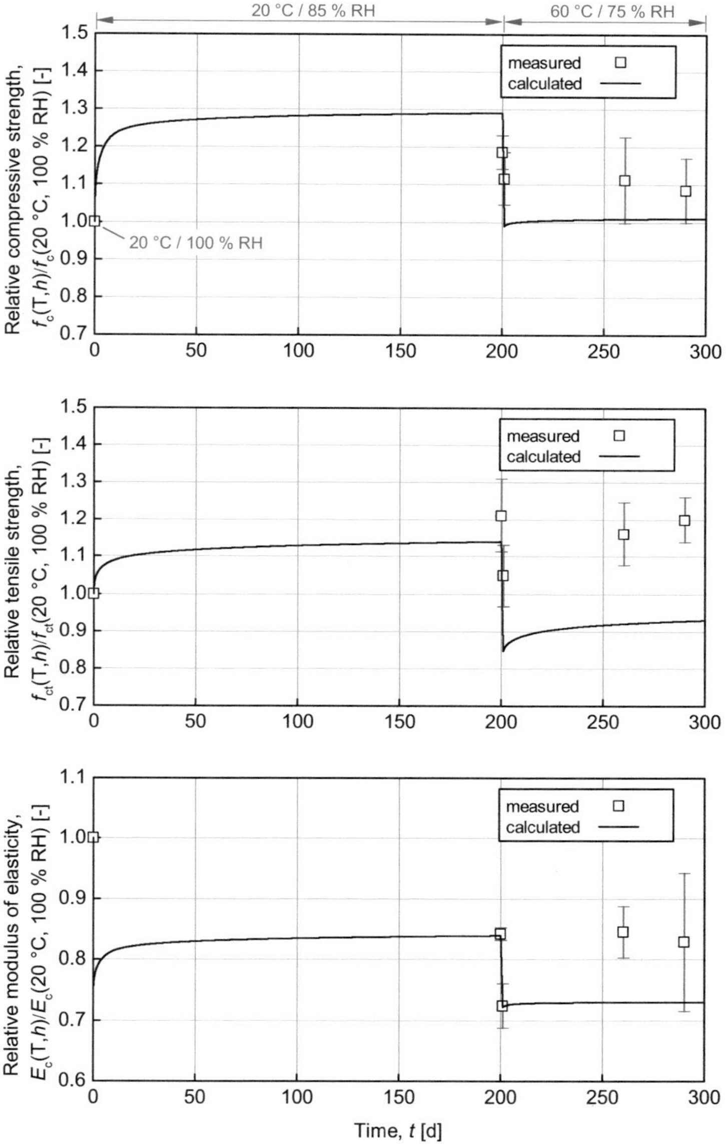

Figure D.5: Comparison between calculations with the material model and measurements of compressive strength (top), tensile strength (middle) and modulus of elasticity (bottom) for the concrete MRC after drying during 200 d at 20 °C / 85 % RH and then heated during 100 d at 60 °C / 75% RH

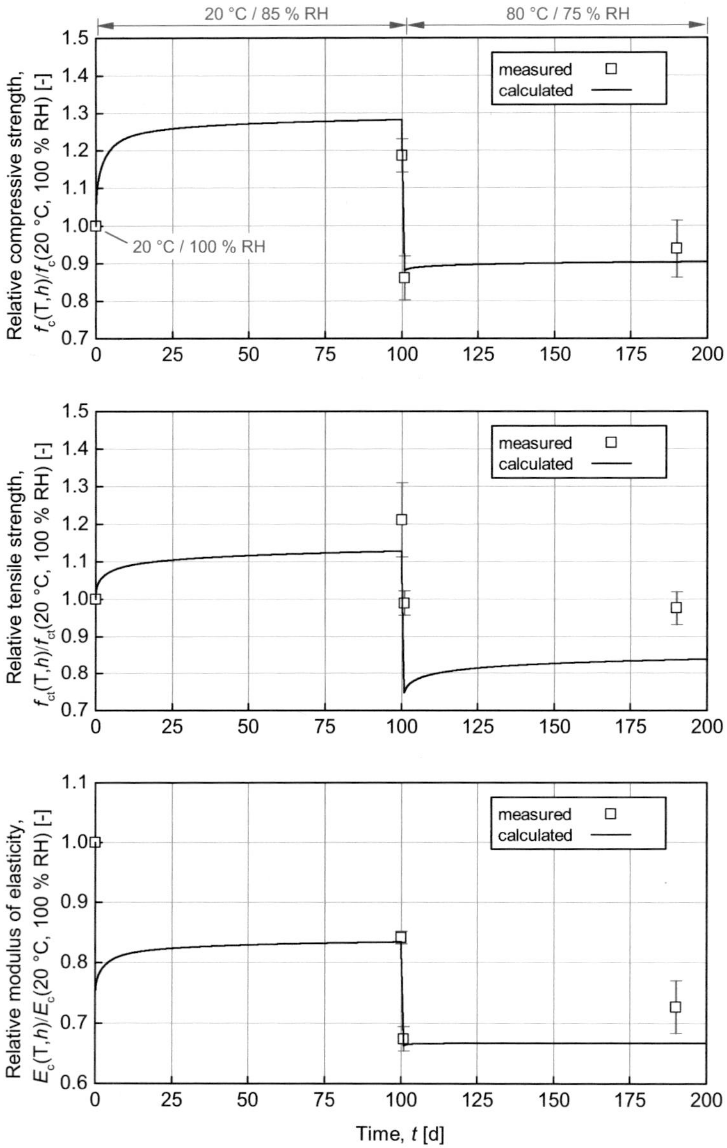

Figure D.6: Comparison between calculations with the material model and measurements of compressive strength (top), tensile strength (middle) and modulus of elasticity (bottom) for the concrete MRC after drying during 100 d at 20 °C / 85 % RH and then heated during 100 d at 80 °C / 75% RH

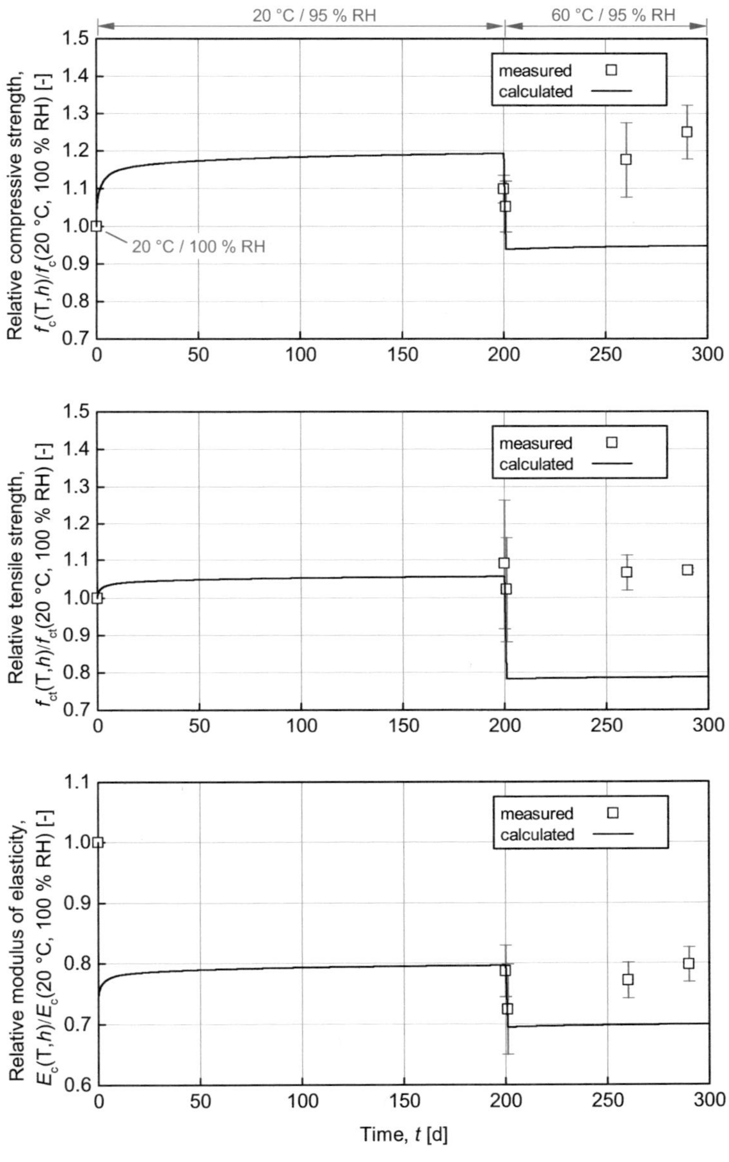

Figure D.7: Comparison between calculations with the material model and measurements of compressive strength (top), tensile strength (middle) and modulus of elasticity (bottom) for the concrete MRC after drying during 200 d at 20 °C / 95 % RH and then heated during 100 d at 60 °C / 95% RH

D.3 Concrete shrinkage

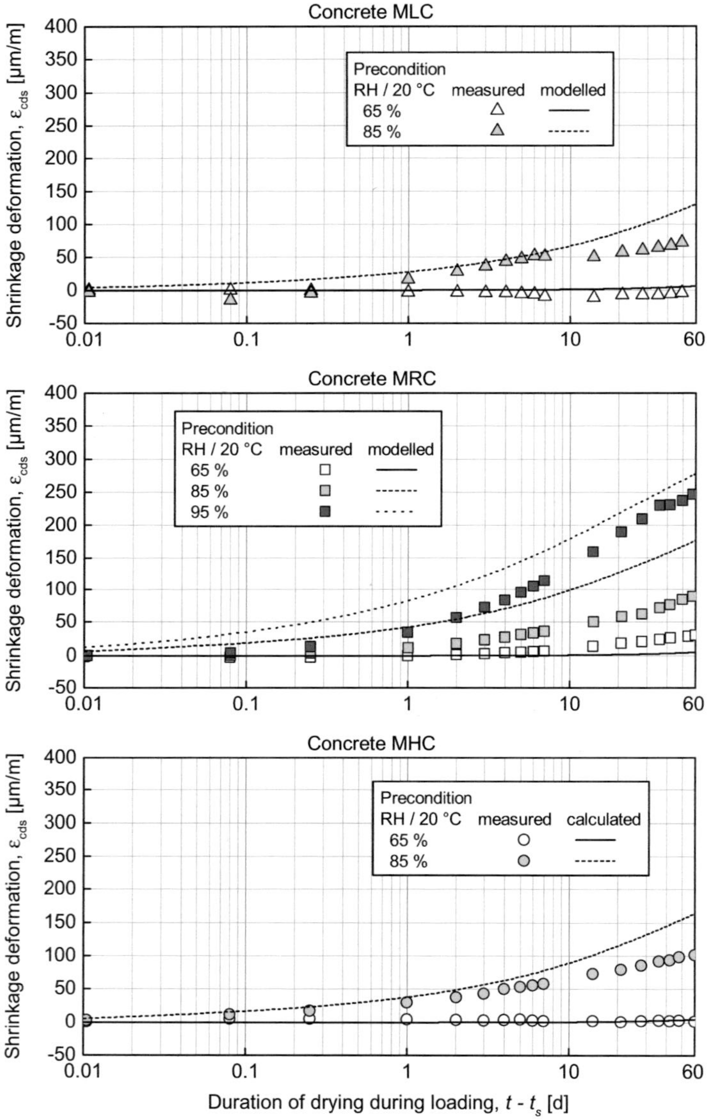

Figure D.8: Comparison between calculations with the material model and measurements of concrete shrinkage for the concretes MLC (top), MRC (middle) MHC (bottom) drying at 20 °C and 65 % RH

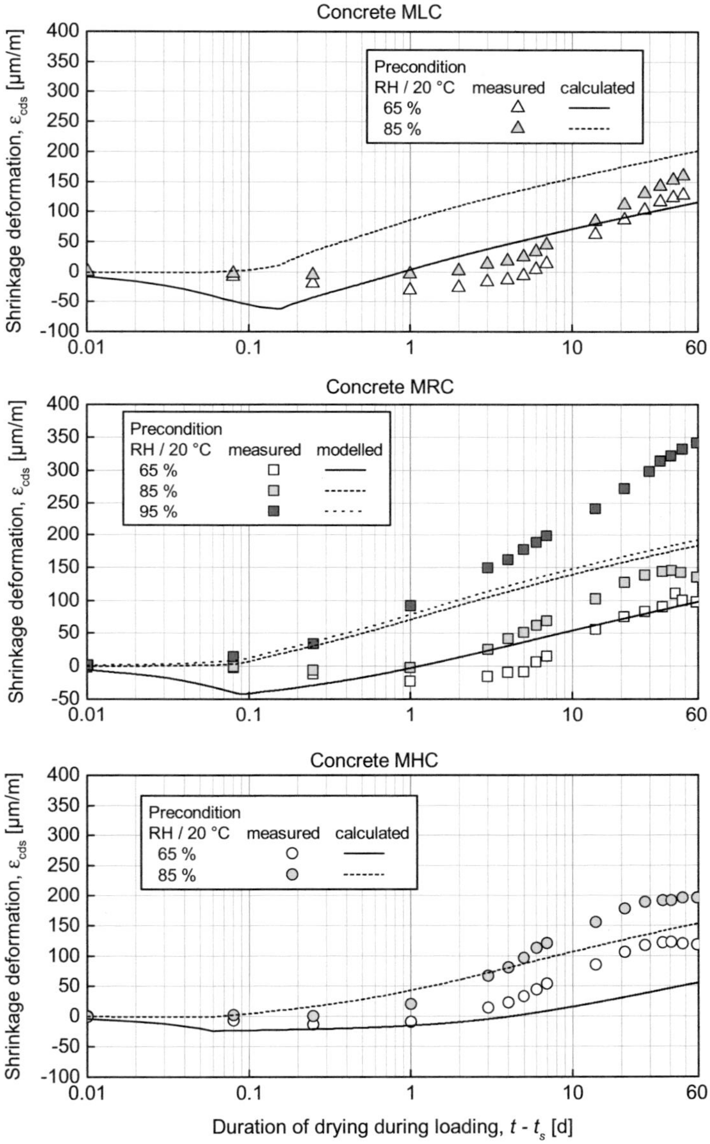

Figure D.9: Comparison between calculations with the material model and measurements of concrete shrinkage for the concretes MLC (top), MRC (middle) MHC (bottom) drying at 70 °C and 65 % RH

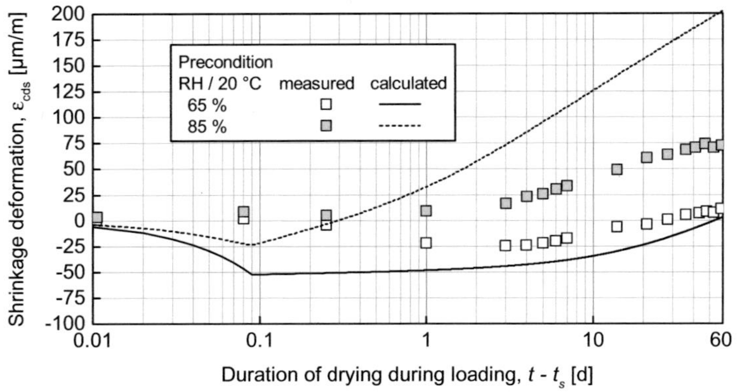

Figure D.10: Comparison between calculations with the material model and measurements of concrete shrinkage for the concretes MRC drying at 40 °C and 65 % RH

D.4 Concrete creep

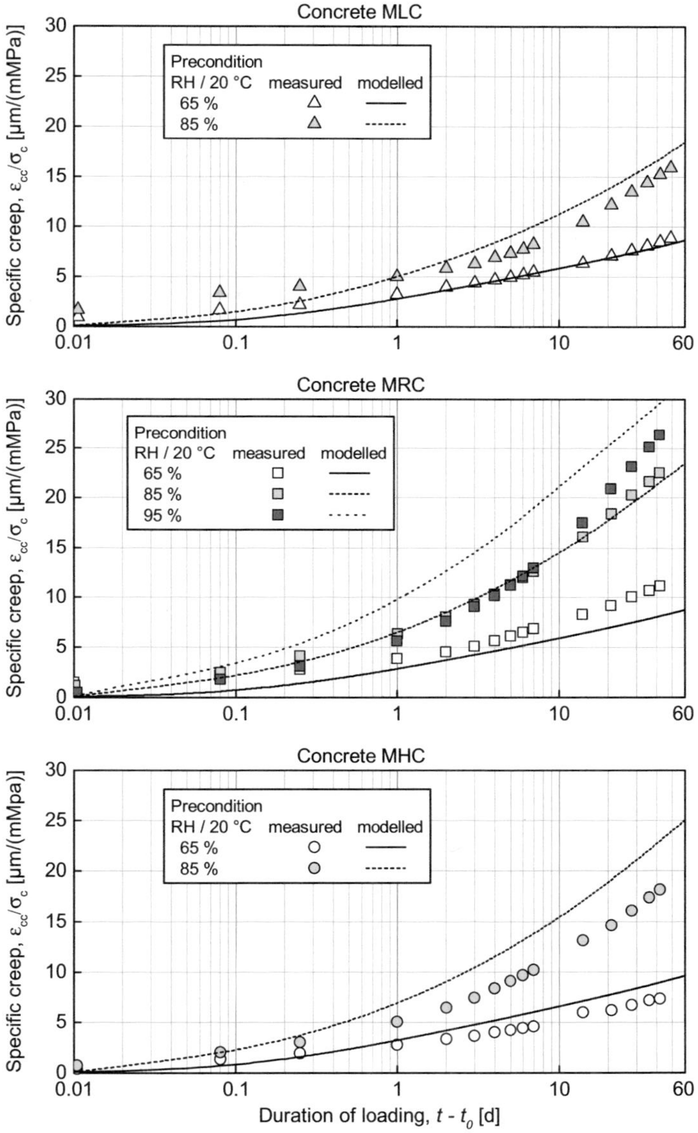

Figure D.11: Specific creep calculated with the material model in comparison with the measured values for the concretes MLC (top), MRC (middle) and MHC (bottom) during the creep tests conducted at 20 °C and 65 % RH

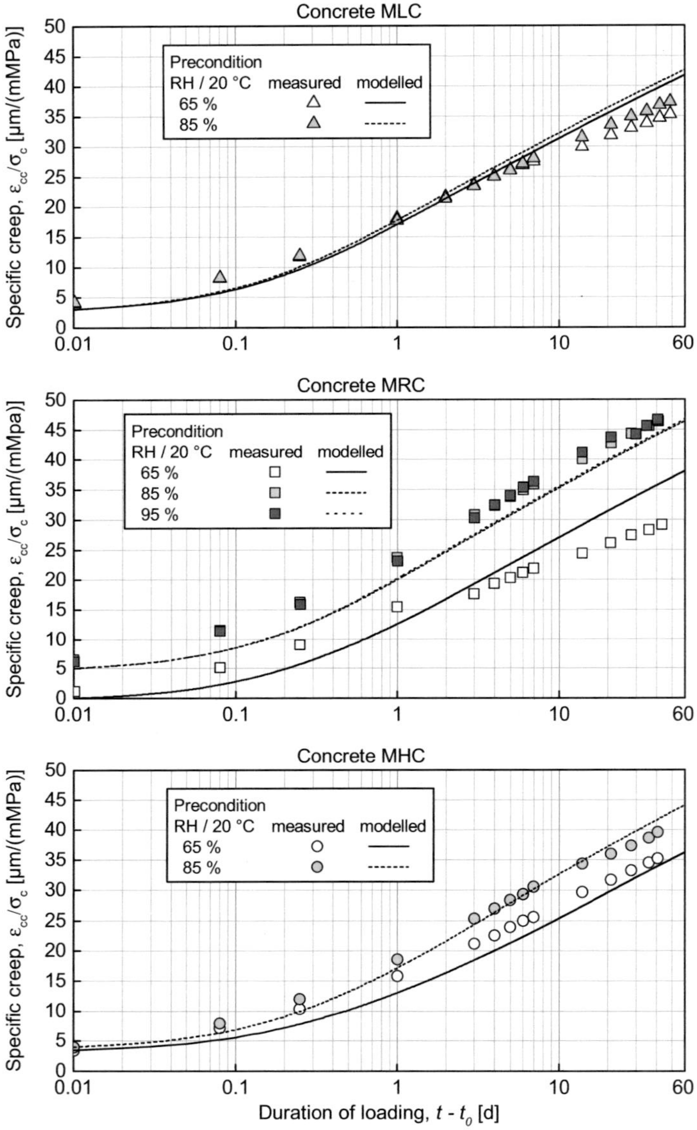

Figure D.12: Specific creep calculated with the material model in comparison with the measured values for the concretes MLC, MRC and MHC during the creep tests conducted at 70 °C and 65 % RH

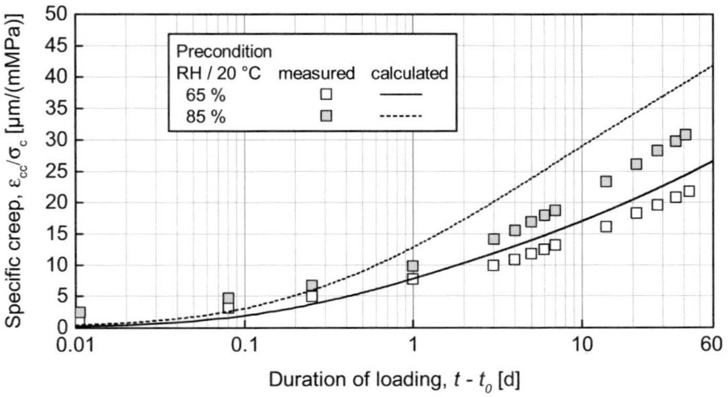

Figure D.13: Specific creep calculated with the material model in comparison with the measured values for the concretes MRC during the creep tests conducted at 40 °C and 65 % RH

Appendix E

Equations of the developed material model

This appendix contains the equations required for the implementation of the material model. The equations are grouped according to the components of the model: moisture transport, strength and stiffness, drying shrinkage and creep.

E.1 Model of moisture transport

Fick's second law of diffusion

$$\frac{\partial h(r,t)}{\partial t} = \frac{1}{r} \cdot \frac{\partial}{\partial r} \left(D(h,T) \cdot r \cdot \frac{\partial h(r,t)}{\partial r} \right)$$

Dependencies of the diffusion coefficient on the relative humidity, the temperature and the w/c-ratio of the concrete mixture

$$D(h,T) = D_1 \cdot \mathrm{f}(h, h_\infty, h_0, \frac{dh}{dr}) \cdot \mathrm{g}(T)$$

$$D_1 = 2.2 + \frac{19.8}{1 + (1.83 \cdot (1 - w/c))^{25}}$$

In case of drying:

$$\mathrm{f}(h, h_\infty, \frac{dh}{dr}) = \frac{1}{1 + \left(\frac{1-h}{1 - h_c(h_\infty, dh/dr)} \right)^n}$$

$$n = 6.1 - \frac{1.8}{1 + (2.3 \cdot (1 - w/c))^9}$$

$$h_c(h_\infty, dh/dr) = 0.5 + \frac{0.5}{1 + 2 \cdot (1 - h_\infty)^2} - 0.5 \cdot \left(\left| \frac{dh}{dr} \right| \cdot r_0 \right)^{0.5}$$

In case of wetting:

$$\mathrm{f}(h, h_\infty, h_0, \frac{dh}{dr}) = \frac{1}{1 + \left(\frac{1 - h_\infty + h - h_0}{1 - h_c(h_0, dh/dr)} \right)^n}$$

$$n = 6.1 - \frac{1.8}{1 + (2.3 \cdot (1 - w/c))^9}$$

$$h_c(h_0, dh/dr) = 0.5 + \frac{0.5}{1 + 2 \cdot (1 - h_0)^2} - 0.5 \cdot \left(\left| \frac{dh}{dr} \right| \cdot r_0 \right)^{0.5}$$

$$\mathrm{g}(T) = \frac{T + 273}{293} \cdot \exp \left[\left(\frac{Q}{R_D} \right) \left(\frac{1}{293} - \frac{1}{T + 273} \right) \right]$$

$$\frac{Q}{R_D} = 9000 \cdot \left[0.2 + \frac{0.8}{1 + (2.1 \cdot w/c)^{10}} \right]$$

Hygrothermic coefficient of concrete

$$K(h) = \frac{dh}{dT} = a \cdot (1 - h)^b \cdot \exp\left[c \cdot (1 - h) \right]$$

Parameters to calculate the hygrothermic coefficient according to the w/c-ratio of the concrete mixture

$$a = 400 \cdot (w/c)^{14.3} + 0.15$$
$$b = 12 \cdot (w/c)^{6.5} + 1.15$$
$$c = -900 \cdot (w/c)^{11} - 7.0$$

The relative humidity of the concrete pores due to a change in temperature is calculated as follows:

$$h_T - h_{T_\mathrm{ref}} = K(h_{T_\mathrm{ref}}) \cdot (T - T_\mathrm{ref})$$

E.2 Model of strength and stiffness

Compressive strength

$$f_\mathrm{c}(h, T) = f_\mathrm{c}(h_\mathrm{ref}, T_\mathrm{ref}) \cdot \left(1 + D_{f_\mathrm{c}} + S_{f_\mathrm{c}} \right)$$

Effect of drying

$$D_{f_c} = -0.32 \cdot (h_{T_{ref}})^{26} + 0.32$$

Effect of thermal incompatibilities

$$S_{f_c} = 0.35 \cdot \left(\frac{T-20}{100}\right)^2 + a_{f_c} \cdot \left(\frac{T-20}{100}\right)$$
$$a_{f_c} = 2.3 \cdot 10^{-9} \cdot \exp\left(19.1 \cdot h_{T_{ref}}\right) - 0.55 \cdot \exp\left(0.64 \cdot h_{T_{ref}}\right)$$

Tensile strength

$$f_{ct}(h,T) = f_{ct}(h_{ref}, T_{ref}) \cdot \left(1 + D_{f_{ct}} + S_{f_{ct}}\right)$$

Effect of drying

$$D_{f_{ct}} = -0.44 \cdot (h_{T_{ref}})^4 + 0.44$$

Effect of thermal incompatibilities

$$S_{f_{ct}} = 0.45 \cdot \left(\frac{T-20}{100}\right)^2 + a_{f_{ct}} \cdot \left(\frac{T-20}{100}\right)$$
$$a_{f_{ct}} = 7.6 \cdot 10^{-5} \cdot \exp\left(8.5 \cdot h_{T_{ref}}\right) - 0.72 \cdot \exp\left(0.48 \cdot h_{T_{ref}}\right)$$

Modulus of elasticity

$$E_c(h,T) = E_c(h_{ref}, T_{ref}) \cdot (1 + D_{E_c} + S_{E_c})$$

Effect of drying

$$D_{E_c} = 0.0 \qquad \text{for } h_{T_{ref}} = 1.0$$
$$D_{E_c} = -0.12 \cdot (h_{T_{ref}})^{19} - 0.14 \qquad \text{for } h_{T_{ref}} < 1.0$$

Effect of thermal incompatibilities

$$S_{E_c} = a_{E_c} \cdot \left(\frac{T-20}{100}\right)$$
$$a_{E_c} = 2.3 \cdot 10^{-5} \cdot \exp\left(9.0 \cdot h_{T_{ref}}\right) - 0.2 \cdot \exp\left(0.7 \cdot h_{T_{ref}}\right)$$

E.3 Model of drying shrinkage

$$d\varepsilon_{\text{cds}} = K_{\text{cds}T} \cdot K_{\text{cds}} \cdot dh$$

Influence of temperature and w/c-ratio

$$K_{\text{cds}}(T, w/c) = 1650 \cdot \left[0.78 + \frac{0.22}{1 + (1.76 \cdot w/c)^{14}}\right]$$

$$K_{\text{cds}T} = 1.2 - 0.01 \cdot T$$

E.4 Model of creep

Basic creep coefficient

$$d\varphi_{bc}(t, t_0) = K_{\text{bc}T} \cdot \frac{K_{\text{bc1}}}{(f_{cm})^{0.7}} \cdot \frac{1}{(t - t_0) + K_{\text{bc2}}} \cdot dt$$

$$K_{\text{bc}T} = a_{K_{\text{bc}T}} \cdot \left[b_{K_{\text{bc}T}} + \frac{1 - b_{K_{\text{bc}T}}}{1 + (1.85 \cdot w/c)^{16}}\right]$$

$$a_{K_{\text{bcT}}} = 1 + 2.7 \left(\frac{T - 20}{100}\right)^{0.34}$$

$$b_{K_{\text{bcT}}} = 1.088 - 0.0044 \cdot T$$

$$\text{K}_{bc1} = 1.55 \cdot \left[0.6 + \frac{0.4}{1 + (2 \cdot w/c)^{10}}\right] \cdot h$$

$$\text{K}_{bc2} = 0.2 \cdot h$$

Drying creep coefficient

$$d\varphi_{dc}(t, t_0) = K_{\text{cds}T} \cdot K_{\text{cds}} \cdot K_{\text{dc}} \cdot dh$$

$$K_{\text{dc}} = 3415 \cdot \left[0.88 + \frac{0.12}{1 + (1.85 \cdot w/c)^{11}}\right]$$

Appendix F

Matlab scripts

F.1 Calculation of moisture transport at constant conditions of temperature and humidity

The following script calculates the development of the relative humidity within concrete member with the shape of a cylinder. Some information must be provided as input in order to run the model. The length of the elements over the radius dr, the radius of the cylinder R, the total time of calculation t, the w/c-ratio of the concrete mixture wc, the initial relative humidity of the concrete pores H0, which can be either a constant or a vector, the environmental relative humidity Hinf and the temperature T. All dimensions must be given in mm, the time in days, the temperature in °C, and the relative humidity in values between 0 and 1.0.

```
function [ ] = Diffusion( dr,R,t,wc,H0,Hinf,T)
% Calculates the water transport with variable diffusion coefficient

%% Definition of variables
  I=R/dr+1;                          % Number of nodes over the geometry
  J=2000;                            % Number of time increments

%% Calculation of the parameters D1 and n
  D1 = 2.2+19.8/(1+(1.83*(1-wc))^25);
  n = 6.1-1.8/(1+(2.3*(1-wc))^9);

%% Initiation of vectors
  HH=zeros(I,1);
  HH=HH+H0;                          % Vector of pore relative humidity
  HH2=HH;
  HHp=zeros(I,1);
  HHn=zeros(I,1);
  Di=zeros(I,1)+D1;                  % Vector of the diffusion coefficient
  Dip=zeros(I,1)+D1;
  Din=zeros(I,1)+D1;
  A=zeros(I,1);
  B=zeros(I,1);
```

```
  C=zeros(I,1);                        % A, B and C initiate with 0.0
  r=zeros(I,1);                        % Vector of the radius
  for k=1:I
   r(k)=dr*(k-1);
  end
  ttot=zeros(J,1);                     % Vector of the total time
  ttot(2)=0.0001;
  dtt=1/1983.137451*log(t/0.000106);   % Time increment
  for i=3:J
   ttot(i)=ttot(i-1)+ttot(i-1)*dtt;
  end

% Influence of temperature
  Q=9000*(0.2+0.8/(1+(2.1*wc)^10));
  D1=D1*(T+273)/(273+20)*exp(Q*(1/(273+20)-1/(273+T)));

%% Calculation of the relative humidity
  for j=2:J
   dt=ttot(j)*dtt;
   for j2=1:10                         % Number of iteration within a dt
    for k=2:I
     HHp(k-1)=HH(k);
     HHn(k)=HH(k-1);
    end
    HHn(1)=HH(1);
    HHp(I)=HH(I);

%% Calculating the moisture dependent diffusion coefficient D for drying
    Di=D1*(1./(1+((1-HH)./(1-(-0.5*(abs(-HHp+HHn)/(2*dr)).^(1/2)...
      +(0.5+0.5/((1+2*(1-Hinf)^2)))))).^n));
    Dip=D1*(1./(1+((1-HHp)./(1-(-0.5*(abs(-HHp+HH)/(dr)).^(1/2)...
      +(0.5+0.5/((1+2*(1-Hinf)^2)))))).^n));
    Din=D1*(1./(1+((1-HHn)./(1-(-0.5*(abs(-HH+HHn)/(dr)).^(1/2)...
      +(0.5+0.5/((1+2*(1-Hinf)^2)))))).^n));

%% Calculating the moisture dependent diffusion coefficient D for wetting
    %Di=D1*(1./(1+((1-Hinf+HH-H0)./(1-(-0.5*(abs(-HHp+HHn)/(2*dr)). ^(1/2)...
      %+(0.5+0.5/((1+2*(1-H0)^2)))))).^n));
    %Dip=D1*(1./(1+((1-Hinf+HHp-H0)./(1-(-0.5*(abs(-HHp+HH)/(dr)).^(1/2)...
      %+(0.5+0.5/((1+2*(1-H0)^2)))))).^n));
    %Din=D1*(1./(1+((1-Hinf+HHn-H0)./(1-(-0.5*(abs(-HH+HHn)/(dr)).^(1/2)...
      %+(0.5+0.5/((1+2*(1-H0)^2)))))).^n));
```

```
%% Building the Matrix
    A=Di./(2*r*dr)*dt;
    B=Di/(dr^2)*dt;
    C=(Dip-Din)/(4*dr^2)*dt;

% Redefining matrix of constants
    M=zeros(I);
    M(1,1)=1+2*B(1);
    M(1,2)=-2*B(1);

    for k=2:I-1
     M(k,k-1)=A(k)-B(k)+C(k);
     M(k,k)=1+2*B(k);
     M(k,k+1)=-A(k)-B(k)-C(k);
    end

% Solving to find new values of relative humidity
    M(I,I)=1.0;
    HH2(I)=Hinf;
    HH=M\HH2;
    end
   HH2=HH;
   end
  end
```

F.2 Program to calculate the influence of temperature and moisture content on the mechanical properties of concrete

Main routine: Tempecon

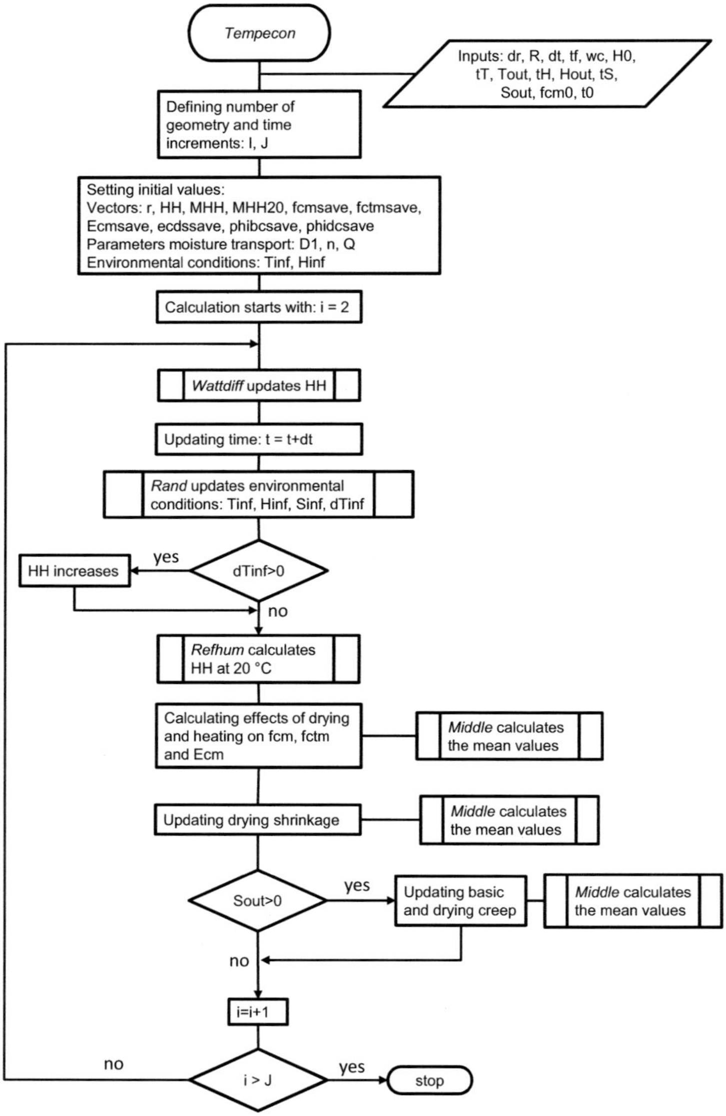

Figure F.1: Flow chart of the main routine Tempecon

```
function [ ] = Tempecon(dr,R,dt,tf,wc,H0,tT,Tout,tH,Hout,tS,Sout,fcm0,t0)
% Effect of temperature on concrete
% Describes the influences of varying the temperature and relative humidity
% of the surroundings on the mechanical properties of concrete:
% Compressive strength, tensile strength, modulus of elasticity, creep and shrinkage

%% Inputs
% dr                          % Increment of the radius [mm]
% R                           % Radius of the sample [mm]
% dt                          % Time increment [d]
% tf                          % Final time of the calculation [d]
% wc                          % w/c-ratio of the concrete [-]
% H0                          % Initial relative humidity (it can be whether
                              % a unique value or a vector) [-]
% tT                          % Vector of time to describe the time
                              %development of the temperature [d]
% Tout                        % Vector of temperature to describe the time
                              % development of temperature [°C]
% tH                          % Vector of time to describe the time
                              % development of the rel. humidity [d]
% Hout                        % Vector of rel. humidity to describe
                              % the time development of the
                              % rel. humidity [-]
% tS                          % Vector of time to describe the time
                              % development of the Load [d]
% Sout                        % Vector of loading to define the interval
                              % in which the sample is loaded [-]
% fcm0                        % Concrete compressive strength at
                              % reference conditions [N/mm^2]
% t0                          % Time at which the load is applied [d]

%% Definition of variables
  I=R/dr+1;                   % Number of nodes over the geometry
  J=round(tf/dt+1);           % Number of time increments
  HH=zeros(I,1);              % Vector of relative humidity
  MHH=zeros(I,J);             % MHH contains the values of HH at
                              % all time increments
  MHH20=zeros(I,J);           % MHH20 contains the values of HH at
                              % ref. temperature and all time increments
  fcmsave=zeros(1,J);         % Vector of fcm
  fctmsave=zeros(1,J);        % Vector of fctm
  Ecmsave=zeros(1,J);         % Vector of Ecm
  ecdssave=zeros(1,J);        % Vector of drying shrinkage
```

```
  phibcsave=zeros(1,J);              % Vector of basic creep coefficient
  phidcsave=zeros(1,J);              % Vector of drying creep coefficient

  r=zeros(I,1);
  for k=1:I
   r(k)=dr*(k-1);                    % Vector of the radius
  end
  t=0;                               % Total time of the calculation

%% Initiation of vectors
  HH=HH+H0;                          % HH initiates with value H0
  MHH20=zeros(I,J);
  MHH(:,1)=HH;                       % MHH for t=0
  MHH20(:,1)=HH;
  fcmsave(1)=1;
  fctmsave(1)=1;
  Ecmsave(1)=1;

%% Calculation of the parameters D1, n and Q/R
  D1 = 2.2+19.8/(1+(1.83*(1-wc))^25);
  n = 6.1-1.8/(1+(2.3*(1-wc))^9);
  Q=9000*(0.2+0.8/(1+(2.1*wc)^10));

%% Setting the initial conditions of Temperatur and rel. humidity
  Tinf=Tout(1);
  Hinf=Hout(1);
  dTinf=0;

  for i=2:J

% Solving de Differential Equation of Diffusion
   [ HH ] = Watdiff( I,r,dr,dt,D1,n,Hinf,Tinf,HH,Q);
   t=t+dt;

% Calculating the boundaries for the next step
   [Tinf Hinf Sinf dTinf]=Rand(tT,Tout,tH,Hout,tS,Sout,t,Tinf,dt);

% Calculating the effect of a temperature change on the rel. humidity
   if abs(dTinf)>0;
    K=50;
    for q=1:K
     HH=HH.*(1+(400*wc^14.3+0.15)*(1-HH).^(12*wc^6.5+...
        1.15).*exp((-900*wc^11-7.0)*(1-HH))*dTinf/K);
    end
   end
```

```
% Saving results
   MHH(:,i)=HH;

% Calculating the reference rel. humidity at T = 20 °C
   [HH20]=Refhum(HH,Tinf,wc);
   MHH20(:,i)=HH20;

%% Calculating effect of drying on the strength and stiffness
   Dryfcm=-0.30*(HH20).^26+0.32;
   Dryfctm=-0.44*(HH20).^4+0.44;
   DryEcm=-0.12*(HH20).^19-0.14;

%% Calculating effect of heating on the strength and stiffness
   afc=2.3e-9*exp(19.1*HH20)-0.55*exp(0.64*HH20);
   Heatfcm=0.35*((Tinf-20)/100)^2+afc*((Tinf-20)/100);
   afct=7.6e-5*exp(8.5*HH20)-0.72*exp(0.48*HH20);
   Heatfctm=0.45*((Tinf-20)/100)^2+afct*((Tinf-20)/100);
   aEc=2.3e-5*exp(9.0*HH20)-0.2*exp(0.7*HH20);
   HeatEcm=aEc*((Tinf-20)/100);

% Calculating the mechanical properties
   fcm=(1+Dryfcm+Heatfcm);
   fctm=(1+Dryfctm+Heatfctm);
   Ecm=(1+DryEcm+HeatEcm);
% Saving results
   fcmsave(i)=Middle(r,I,fcm);
   fctmsave(i)=Middle(r,I,fctm);
   Ecmsave(i)=Middle(r,I,Ecm);

%% Calculating Drying Shrinkage
   KcdsT=1.2-0.01*Tinf;
   Kcds=1650*(0.78+(0.22)/(1+(1.76*wc)^14));
   ecds=Kcds*KcdsT*(MHH(:,i-1)-HH);
% Saving results
   ecdssave(i)=ecdssave(i-1)+Middle(r,I,ecds);

   if Sinf>0
%% Calculating basic creep
     aKbcT=1+2.7*((Tinf-20)/100)^0.34;
     bKbcT=1.088-0.0044*Tinf;
     KbcT=aKbcT*(bKbcT+(1-bKbcT)/(1+(1.85*wc)^(16)));
     Kbc1=1.55*(0.6+(0.4)/(1+(2*wc)^10))*HH;
     Kbc2=0.2*HH;
     phibc=KbcT*Kbc1/(fcm0)^0.7*1./(t-t0+Kbc2)*dt;
```

```
% Saving results
    phibcsave(i)=phibcsave(i-1)+Middle(r,I,phibc);

%% Calculating drying creep
    Kdc=3415*(0.88+0.12/(1+(1.85*wc)^11));
    phidc=KcdsT*Kcds*Kdc*abs(MHH(:,i-1)-HH)/1000000;
% Saving results
    phidcsave(i)=phidcsave(i-1)+Middle(r,I,phidc);

  end
 end
```

Subroutine: Middle

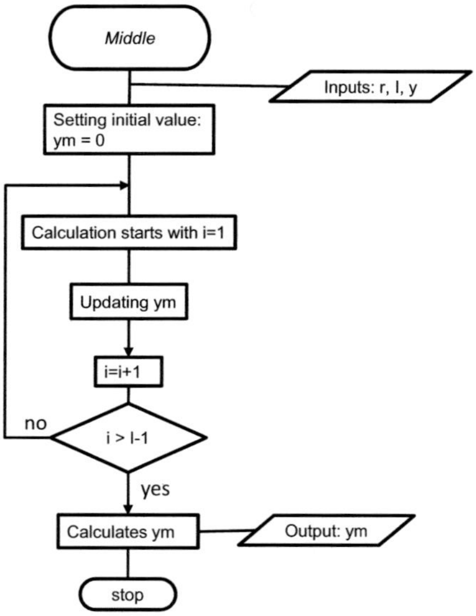

Figure F.2: Flow chart of the subroutine Middle

```
function[ym]=Middle(r,I,y)
% Middle values
% Calculates the middle value over a cylindrical geometry

%% Definitions:
% r                         Vector of the radius
% I                         Number of nodes
% y                         Vector of values to evaluate
% ym                        Mean value
```

```
    ym=0.0;
    for i=1:I-1
     ym=ym+(y(i+1)+y(i))/2*(r(i+1)^2-r(i)^2);
    end

    ym=ym/(r(I)^2-r(1)^2);

end
```

Subroutine: Rand

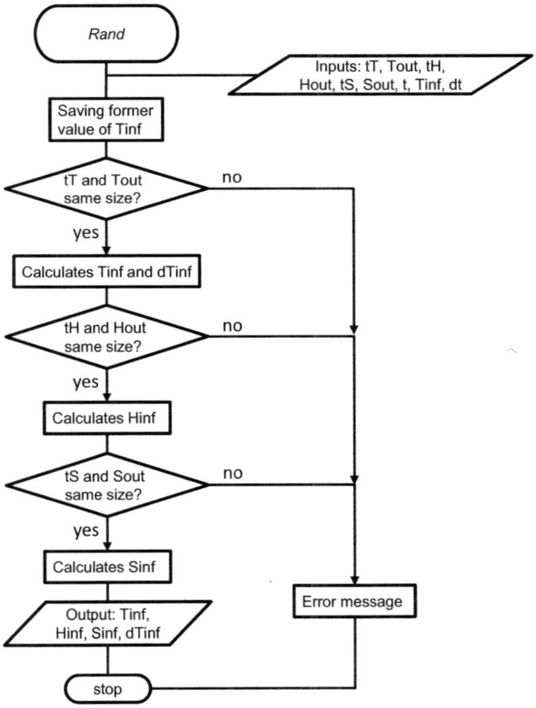

Figure F.3: Flow chart of the subroutine Rand

```
function [ Tinf,Hinf,Sinf,dTinf ] = Rand(tT,Tout,tH,Hout,tS,Sout,t,Tinf,dt)
% Boundary conditions:
% Checks the boundary conditions of the model in every time step.
% Calculates the concrete temperature, relative humidity of the environment
% and creep stress according to the functions given in the input
```

```
%% Definitions:
% tT        vector of time of the temperature function
% Tout      vector of concrete temperature
% tH        vector of time of the function of
            % environmental rel. humidity
% Hout      vector of environmental relative humidity
% tS        vector of time of the creep stress function
% Sout      vector of loading
% t         current time of the calculation
% Tinf      current temperature of the concrete
% dt        time increment
% Hinf      current relative humidity of the environment
% dTinf     changes in the relative humidity of
            % the environment in dt
% Sinf      current load condition

   a=Tinf;

% calculating Tinf
   n=length(tT);
   n2=length(Tout);
   if n~=n2;
    error('length of tT does not match the length of Tout');
   end

   i=1;
   while i<n
    if t<=tT(i+1)+dt/10
     Tinf=(Tout(i+1)-Tout(i))/(tT(i+1)-tT(i))*(t-tT(i))+Tout(i);
     i=n-1;
    end
   i=i+1;
   end
   dTinf=Tinf-a;

% calculating Hinf
   n=length(tH);
   n2=length(Hout);
   if n~=n2;
    error('length of tH does not match the length of Hout');
   end

   i=1;
   while i<n
```

```
    if t<=tH(i+1)+dt/10
     Hinf=(Hout(i+1)-Hout(i))/(tH(i+1)-tH(i))*(t-tH(i))+Hout(i);
     i=n-1;
    end
    i=i+1;
end

% calculating Sinf
   n=length(tS);
   n2=length(Sout);
   if n~=n2;
    error('length of tS does not match the length of Sout');
   end

   i=1;
   while i<n
    if t<=tS(i+1)+dt/10
     Sinf=(Sout(i+1)-Sout(i))/(tS(i+1)-tS(i))*(t-tS(i))+Sout(i);
     i=n-1;
    end
    i=i+1;
   end
  end
```

Subroutine: Refhum

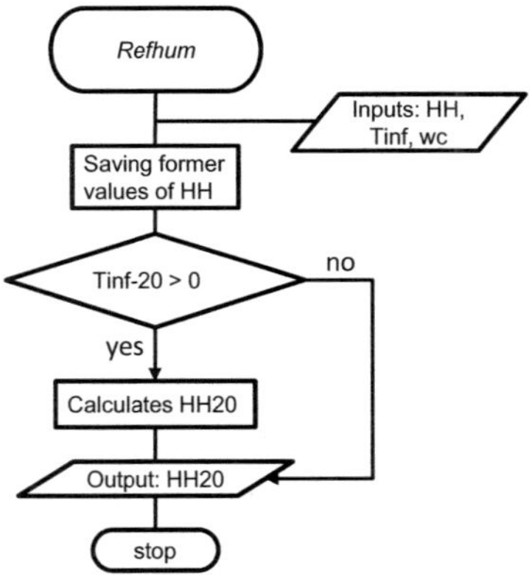

Figure F.4: Flow chart of the subroutine Refhum

```
function[HH20]=Refhum(HH,Tinf,wc)
% Reference relative humidty at 20 °C
% Calculates the rel. humidity of the material at the temperature of reference
```

```
%% Definitions:
% HH                    vector of current relative humidity of the
                        % concrete pores
% Tinf                  current temperature of the concrete
% HH20                  vector of relative humidity of the concrete
                        % equivalent at a temperature of 20 °C

  HH20=HH;
   if abs(Tinf-20)>0;
    K=100;
    for q=1:K
     HH20=HH20.*(1-(400*wc^14.3+0.15)*(1-HH20).^(12*wc^6.5+...
         1.15).*exp((-900*wc^11-7.0)*(1-HH20))*(Tinf-20)/K);
    end
   end
 end
```

Subroutine: Watdiff

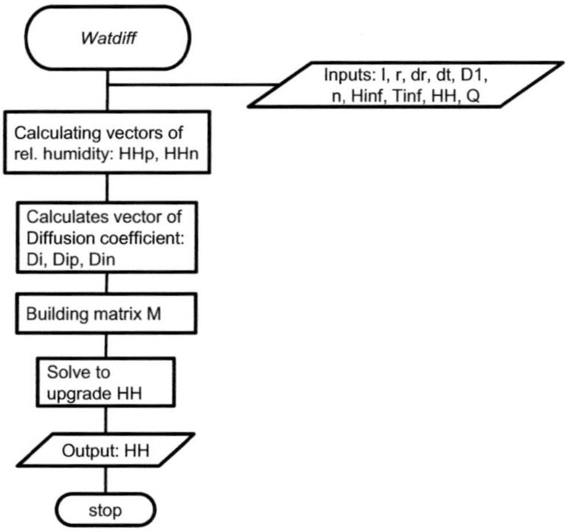

Figure F.5: Flow chart of the subroutine Watdiff

```
function [ HH ] = Watdiff( I,r,dr,dt,D1,n,Hinf,Tinf,HH,Q)
% Calculates the relative humidity of the concrete pores based on the
% Fick's second law

% I                     number of nodes
% r                     vector of radius
% dr                    radial distance between to nodes
```

```
% dt                                time increment
% D1                                Diffusion coefficient at reference conditions
% n                                 exponent for the calculation of the
                                    % diffusion coefficient
% Hinf                              current relative humidity of the environment
% Tinf                              current temperature of the concrete
% HH                                vector of current relative humidity of
                                    % the concrete pores
% Q                                 Quotient between the activation energy of
                                    % diffusion and the universal gas constant

% Initiation of variables
    HHp=zeros(I,1);
    HHn=zeros(I,1);
    C=zeros(I,1);
    A=zeros(I,1);
    B=zeros(I,1);                   % A, B and C initiate with 0.0

    for k=2:I
     HHp(k-1)=HH(k);
     HHn(k)=HH(k-1);
    end

    D1=D1*(Tinf+273)/(273+20)*exp(Q*(1/(273+20)-1/(273+Tinf)));
    Di=D1*(1./(1+((1-HH)./(1-(-0.5*(abs(-HHp+HHn)/(2*dr)).^(1/2)...
    +(0.5+0.5/((1+2*(1-Hinf)^2)))))).^n));
    Dip=D1*(1./(1+((1-HHp)./(1-(-0.5*(abs(-HHp+HH)/(dr)).^(1/2)...
    +(0.5+0.5/((1+2*(1-Hinf)^2)))))).^n));
    Din=D1*(1./(1+((1-HHn)./(1-(-0.5*(abs(-HH+HHn)/(dr)).^(1/2)...
    +(0.5+0.5/((1+2*(1-Hinf)^2)))))).^n));

    A=Di./(2*r*dr)*dt;
    B=Di/(dr^2)*dt;
    C=(Dip-Din)/(4*dr^2)*dt;

% Redefining matrix of constants
    M=zeros(I);
    M(1,1)=1+2*B(1);
    M(1,2)=-2*B(1);

    for k=2:I-1
     M(k,k-1)=A(k)-B(k)+C(k);
     M(k,k)=1+2*B(k);
     M(k,k+1)=-A(k)-B(k)-C(k);
    end
```

```
    M(I,I)=1.0;                    % for free drying
    HH(I)=Hinf;
% M(I,I-1)=-2*B(I);                % for sealed samples
% M(I,I)=1+2*B(I);

    HH=M\HH;
    HH=real(HH);

  end
```

Schriftenreihe des

Instituts für Massivbau und Baustofftechnologie

Herausgeber Prof. Dr.-Ing. Frank Dehn
Prof. Dr.-Ing. Lothar Stempniewski

Institut für Massivbau und Baustofftechnologie
Universität Karlsruhe (TH)
ISSN 0933-0461

Heft 1 **Manfred Curbach**
Festigkeitssteigerung von Beton bei hohen Belastungsgeschwindigkeiten. 1987

Heft 2 **Franz-Hermann Schlüter**
Dicke Stahlbetonplatten unter stoßartiger Belastung – Flugzeugabsturz. 1987

Heft 3 **Marlies Schieferstein**
Der Zugflansch von Stahlbetonplattenbalken unter Längsschub und Querbiegung bei kritischer Druckbeanspruchung von Beton. 1988

Heft 4 **Thomas Bier**
Karbonatisierung und Realkalisierung von Zementstein und Beton. 1988

Heft 5 **Wolfgang Brameshuber**
Bruchmechanische Eigenschaften von jungem Beton. 1988

Heft 6 **Bericht DFG-Forschungsschwerpunkt**
Durability of Non-Metallic Inanorganic Building Materials. 1988

Heft 7 **Manfred Feyerabend**
Der harte Querstoß auf Stützen aus Stahl und Stahlbeton.1988

Heft 8 **Klaus F. Schönlin**
Permeabilität als Kennwert der Dauerhaftigkeit von Beton. 1989

Heft 9 **Lothar Stempniewski**
Flüssigkeitsgefüllte Stahlbetonbehälter unter Erdbebeneinwirkung.1990

Heft 10 **Jörg Weidner**
Vergleich von Stoffgesetzen granularer Schüttgüter zur Silodruckermittlung. 1990

Heft 11 **Pingli Yi**
Explosionseinwirkungen auf Stahlbetonplatten. 1991

Schriftenreihe des

Instituts für Massivbau und Baustofftechnologie

Heft 12	**Rainer Kunterding** Beanspruchung der Oberfläche von Stahlbetonsilos durch Schüttgüter. 1991
Heft 13	**Peter Haardt** Zementgebundene und kunststoffvergütete Beschichtungen auf Beton. 1991
Heft 14	**Günter Rombach** Schüttguteinwirkungen auf Silozellen – Exzentrische Entleerung. 1991
Heft 15	**Harald Garrecht** Porenstrukturmodelle für den Feuchtehaushalt von Baustoffen mit und ohne Salzbefrachtung und rechnerische Anwendung auf Mauerwerk. 1992
Heft 16	**Violandi Vratsanou** Das nichtlineare Verhalten unbewehrter Mauerwerksscheiben unter Erdbebenbeanspruchung – Hilfsmittel zur Bestimmung der q-Faktoren. 1992
Heft 17	**Carlos Rebelo** Stochastische Modellierung menschenerzeugter Schwingungen. 1992
Heft 18	**Seminar 29./30. März 1993** Erdbebenauslegung von Massivbauten unter Berücksichtigung des Eurocode 8. 1993
Heft 19	**Hubert Bachmann** Die Massenträgheit in einem Pseudo-Stoffgesetz für Beton bei schneller Zugbeanspruchung. 1993
Heft 20	**DBV/AiF-Forschungsbericht H. Emrich** Zum Tragverhalten von Stahlbetonbauteilen unter Querkraft- und Längszugbeanspruchung. 1993
Heft 21	**Robert Stolze** Zum Tragverhalten von Stahlbetonplatten mit von den Bruchlinien abweichender Bewehrungsrichtung – Bruchlinien-Rotationskapazität. 1993
Heft 22	**Jie Huang** Extern vorgespannte Segmentbrücken unter kombinierter Beanspruchung aus Biegung, Querkraft und Torsion. 1994

Schriftenreihe des

Instituts für Massivbau und Baustofftechnologie

Heft 23 **Rolf Wörner**
Verstärkung von Stahlbetonbauteilen mit Spritzbeton. 1994

Heft 24 **Ioannis Retzepis**
Schiefe Betonplatten im gerissenen Zustand. 1995

Heft 25 **Frank Dahlhaus**
Stochastische Untersuchungen von Silobeanspruchungen. 1995

Heft 26 **Cornelius Ruckenbrod**
Statische und dynamische Phänomene bei der Entleerung von Silozellen. 1995

Heft 27 **Shishan Zheng**
Beton bei variierender Dehngeschwindigkeit, untersucht mit einer neuen modifizierten Split-Hopkinson-Bar-Technik. 1996

Heft 28 **Yong-zhi Lin**
Tragverhalten von Stahlfaserbeton. 1996

Heft 29 **DFG**
Korrosion nichtmetallischer anorganischer Werkstoffe im Bauwesen. 1996

Heft 30 **Jürgen Ockert**
Ein Stoffgesetz für die Schockwellenausbreitung in Beton. 1997

Heft 31 **Andreas Braun**
Schüttgutbeanspruchungen von Silozellen unter Erdbebeneinwirkung. 1997

Heft 32 **Martin Günter**
Beanspruchung und Beanspruchbarkeit des Verbundes zwischen Polymerbeschichtungen und Beton. 1997

Heft 33 **Gerhard Lohrmann**
Faserbeton unter hoher Dehngeschwindigkeit. 1998

Heft 34 **Klaus Idda**
Verbundverhalten von Betonrippenstäben bei Querzug. 1999

Heft 35 **Stephan Kranz**
Lokale Schwind- und Temperaturgradienten in bewehrten, oberflächennahen Zonen von Betonstrukturen. 1999

Schriftenreihe des

Instituts für Massivbau und Baustofftechnologie

Heft 36 **Gunther Herold**
Korrosion zementgebundener Werkstoffe in mineralsauren Wässern. 1999

Heft 37 **Mostafa Mehrafza**
Entleerungsdrücke in Massefluss-Silos – Einflüsse der Geometrie und Randbedingungen. 2000

Heft 38 **Tarek Nasr**
Druckentlastung bei Staubexplosionen in Siloanlagen. 2000

Heft 39 **Jan Akkermann**
Rotationsverhalten von Stahlbeton-Rahmenecken. 2000

Heft 40 **Viktor Mechtcherine**
Bruchmechanische und fraktologische Untersuchungen zur Rißausbreitung in Beton. 2001

Heft 41 **Ulrich Häußler-Combe**
Elementfreie Galerkin-Verfahren – Grundlagen und Einsatzmöglichkeiten zur Berechnung von Stahlbetontragwerken. 2001

Heft 42 **Björn Schmidt-Hurtienne**
Ein dreiaxiales Schädigungsmodell für Beton unter Einschluß des Dehnrateneffekts bei Hochgeschwindigkeitsbelastung. 2001

Heft 43 **Nazir Abdou**
Ein stochastisches nichtlineares Berechnungsverfahren für Stahlbeton mit finiten Elementen. 2002

Heft 44 **Andreas Plokitza**
Ein Verfahren zur numerischen Simulation von Betonstrukturen beim Abbruch durch Sprengen. 2002

Heft 45 **Timon Rabczuk**
Numerische Untersuchungen zum Fragmentierungsverhalten von Beton mit Hilfe der SPH-Methode. 2002

Heft 46 **Norbert J. Krutzik**
Zu Anwendungsgrenzen von FE-Modellen bei der Simulation von Erschütterungen in Kernkraftbauwerken bei Stoßbelastungen. 2002

Heft 47 **Thorsten Timm**
Beschuß von flüssigkeitsgefüllten Stahlbehältern. 2002

Schriftenreihe des
Instituts für Massivbau und Baustofftechnologie

Heft 48	**Slobodan Kasic** Tragverhalten von Segmentbauteilen mit interner und externer Vorspannung ohne Verbund. 2002
Heft 49	**Christoph Kessler-Kramer** Zugtragverhalten von Beton unter Ermüdungsbeanspruchung. 2002
Heft 50	**Nico Herrmann** Experimentelle Verifizierung von Prognosen zur Sprengtechnik. 2002
Heft 51	**Michael Baur** Elastomerlager und nichtlineare Standorteffekte bei Erdbebeneinwirkung. 2003
Heft 52	**Seminar 02. Juli 2004** DIN 1045-1; Aus der Praxis für die Praxis. 2004
Heft 53	**Abdelkhalek Saber Omar Mohamed** Behaviour of Retrofitted Masonry Shear Walls Subjected to Cyclic Loading. 2004
Heft 54	**Werner Hörenbaum** Verwitterungsmechanismen und Dauerhaftigkeit von Sandsteinsichtmauenwerk. 2005
Heft 55	**Seminar Februar 2006** DIN 4149 – Aus der Praxis für die Praxis. 2006
Heft 56	**Sam Foos** Unbewehrte Betonfahrbahnplatten unter witterungsbedingten Beanspruchungen. 2006
Heft 57	**Ramzi Maliha** Untersuchungen zur Rissbildung in Fahrbahndecken aus Beton. 2006
Heft 58	**Andreas Fäcke** Numerische Simulation des Schädigungsverhaltens von Brückenpfeilern aus Stahlbeton unter Erdbebenlasten. 2006
Heft 59	**Juliane Möller** Rotationsverhalten von verbundlos vorgespannten Segmenttragwerken. 2006

Schriftenreihe des
Instituts für Massivbau und Baustofftechnologie

Heft 60 **Martin Larcher**
Numerische Simulation des Betonverhaltens unter Stoßwellen mit Hilfe des Elementfreien Galerkin-Verfahrens. 2007

Heft 61 **Christoph Niklasch**
Numerische Untersuchungen zum Leckageverhalten von gerissenen Stahlbetonwänden. 2007

Heft 62 **Halim Khbeis**
Experimentelle und numerische Untersuchungen von Topflagern. 2007

Heft 63 **Sascha Schnepf**
Vereinfachte numerische Simulation des Tragverhaltens ebener mauerwerksausgefachter Stahlbetonrahmen unter zyklischer Belastung. 2007

Heft 64 **Christian Wallner**
Erdbebengerechtes Verstärken von Mauerwerk durch Faserverbundwerkstoffe – experimentelle und numerische Untersuchungen. 2008

Heft 65 **Niklas Puttendörfer**
Ein Beitrag zum Gleitverhalten und zur Sattelausbildung externer Spannglieder. 2008

»»»»»»»»» **Bezug der Hefte 1 – 65 und 67**
Institut für Massivbau und Baustofftechnologie
Karlsruher Institut für Technologie (KIT)
Gotthard-Franz-Str. 3, 76131 Karlsruhe
www.betoninstitut.de

Bezug ab Heft 66
KIT Scientific Publishing
Straße am Forum 2, 76131 Karlsruhe
www.ksp.kit.edu

»»»»»»»»» **Fortführung der Reihe ab Heft 66 unter neuem Namen**

KARLSRUHER REIHE
Massivbau
Baustofftechnologie
Materialprüfung

erschienen bei KIT Scientific Publishing (ISSN 1869-912X)

KARLSRUHER REIHE

Massivbau – Baustofftechnologie – Materialprüfung

Herausgeber Prof. Dr.-Ing. Frank Dehn
Prof. Dr.-Ing. Lothar Stempniewski

Institut für Massivbau und Baustofftechnologie
Materialprüfungs- und Forschungsanstalt, MPA Karlsruhe

Karlsruher Institut für Technologie (KIT)
KIT Scientific Publishing
ISSN 1869-912X

Heft 66 **Michael Haist**
Zur Rheologie und den physikalischen Wechselwirkungen bei Zementsuspensionen. 2009
ISBN 978-3-86644-475-1

Heft 67 **Stephan Steiner**
Beton unter Kontaktdetonation – neue experimentelle Methoden. 2009
(noch erschienen in der Schriftenreihe des Instituts für Massivbau und Baustofftechnologie, ISSN 0933-0461)

Heft 68 **Christian Münich**
Hybride Multidirektionaltextilien zur Erdbebenverstärkung von Mauerwerk – Experimente und numerische Untersuchungen mittels eines erweiterten Makromodells. 2011
ISBN 978-3-86644-734-9

Heft 69 **Viktória Malárics**
Ermittlung der Betonzugfestigkeit aus dem Spaltzugversuch an zylindrischen Betonproben. 2011
ISBN 978-3-86644-735-6

Heft 70 **Daniela Ruch**
Bestimmung der Last-Zeit-Funktion beim Aufprall flüssigkeitsgefüllter Stoßkörper. 2011
ISBN 978-3-86644-736-3

Heft 71 **Marc Beitzel**
Frischbetondruck unter Berücksichtigung der rheologischen Eigenschaften. 2012
ISBN 978-3-86644-783-7

Die Bände sind unter www.ksp.kit.edu als PDF frei verfügbar oder als Druckausgabe bestellbar.

KARLSRUHER REIHE

Massivbau – Baustofftechnologie – Materialprüfung

Heft 72 **Michael Stegemann**
Großversuche zum Leckageverhalten von gerissenen Stahlbetonwänden. 2012
ISBN 978-3-86644-860-5

Heft 73 **Isabel Anders**
Stoffgesetz zur Beschreibung des Kriech- und Relaxationsverhaltens junger normal- und hochfester Betone. 2013
ISBN 978-3-7315-0043-8

Heft 74 **Jennifer C. Scheydt**
Mechanismen der Korrosion bei ultrahochfestem Beton. 2013
ISBN 978-3-7315-0113-8

Heft 75 **Michael Auer**
Ein Verbundmodell für Stahlbeton unter Berücksichtigung der Betonschädigung. 2015
ISBN 978-3-7315-0316-3

Heft 76 **Christian Moritz Urban**
Experimentelle Untersuchungen und Bemessungsansätze für faserverstärktes Mauerwerk unter Erdbebenbeanspruchungen. 2015
ISBN 978-3-7315-0372-9

Heft 77 **Tobias Bacht**
Horizontaltragfähigkeit von Wänden aus Leichtbeton-Schalungssteinen – Experimente und numerische Modellierung. 2015
ISBN 978-3-7315-0413-9

Heft 78 **Björn Haag**
Schadensidentifikation mit modalen Parametern: Anwendung auf extern vorgespannte Hohlkastenbrücken. 2016
ISBN 978-3-7315-0458-0

Heft 79 **Engin Kotan**
Ein Prognosemodell für die Verwitterung von Sandstein. 2017
ISBN 978-3-7315-0520-4

Heft 80 **Vladislav Kvitsel**
Zur Vorhersage des Schwindens und Kriechens von normal- und hochfestem Konstruktionsleichtbeton mit Blähtongesteinskörnung. 2017
ISBN 978-3-7315-0521-1

Die Bände sind unter www.ksp.kit.edu als PDF frei verfügbar oder als Druckausgabe bestellbar.

KARLSRUHER REIHE

Massivbau – Baustofftechnologie – Materialprüfung

Heft 81 **Michael Vogel**
Schädigungsmodell für die Hydroabrasionsbeanspruchung zur probabilistischen Lebensdauerprognose von Betonoberflächen im Wasserbau. 2017
ISBN 978-3-7315-0522-8

Heft 82 **Georgios Maltidis**
Seismic soil structure interaction of navigation locks. 2017
ISBN 978-3-7315-0718-5

Heft 83 **Steffen Siegel**
Zustandsbestimmung von externen Spanngliedern und Schrägseilen mit Frequenzanalysen. 2018
ISBN 978-3-7315-0772-7

Heft 84 **Fernando Acosta Urrea**
Influence of elevated temperatures up to 100 °C on the mechanical properties of concrete. 2018
ISBN 978-3-7315-0795-6